Current Topics in Microbiology 162 and Immunology

Editors

Bluetongue Viruses

Edited by
P. Roy and B. M. Gorman

With 37 Figures

Springer-Verlag Berlin Heidelberg GmbH

Polly Roy MSc, Ph.D.
Dept. of Environmental Health Sciences,
School of Public Health,
University of Alabama at Birmingham,
Birmingham, AL 35294, USA

Barry M. Gorman MSc, Ph.D.
United States Dept. of Agriculture,
Agricultural Research Service,
Arthropod-borne Animal Diseases
Research Laboratory,
Laramie, WY82071-3965, USA

ISBN 978-3-642-75249-0 **ISBN 978-3-642-75247-6 (eBook)**
DOI 10.1007/978-3-642-75247-6

Softcover reprint of the hardcover 1st edition 1990
Library of Congress Catalog Card Number 15-12910

Typesetting: Thomson Press (India) Ltd., New Delhi
Offsetprinting: Saladruck, Berlin; Bookbinding: B. Helm Berlin
2123/3020-543210 – Printed on acid-free paper

Preface

Bluetongue viruses (BTV) cause diseases that have serious economic consequences in ruminants (sheep, cattle) in many parts of the world. The incidence of bluetongue disease affects the international movement of animals and germ plasm. Although the etiological agent of the disease was isolated in 1900 and preliminary biochemical characterizations were published as early as in 1969, most of the current understanding of the molecular biology, biochemistry, and genetics of BTV has evolved only recently. Triggered by the modern techniques of molecular biology, genetics, and immunology, BTV research has experienced an information explosion in the past 10 years. However, much of this information is scattered throughout an extensive literature. It is therefore an appropriate time to meld this together into a reference book. This book includes comprehensive information on BTV research provided in articles contributed by researchers from around the world. It covers what is known about the molecular structure of the virus and the current understanding of its biology, evolution, and relationships with its invertebrate and vertebrate hosts (infection, immunity, and pathogenicity).

Specific topics covered include a short description of the emergence of the disease and its infection patterns. Diagnosis of the disease, serology of the infection, pathology, and pathogenesis are described, in addition to the immune response to infection in the natural host and in model animals. Much progress has been made recently in understanding BTV replication from studies of in vitro culture systems. This is discussed with detailed descriptions of how immunoelectron microscopy has contributed to our understanding of the various steps in virus morphogenesis and the release of progeny viruses from infected cells. BTV is transmitted by *Culicoides* vectors to various vertebrate hosts. A separate chapter is therefore devoted to the BTV vector relationships.

Inevitably molecular and genetic studies have come to dominate BTV research in recent years, and this is reflected in at least three chapters. The importance of epidemiology, evolution of multiple BTV serotypes and their relationships with each other has been appreciated for many years. Only recently has this been put on to a fundamental basis. The new information on BTV regarding the structure and function of the various viral genes and their protein products has clarified many of the puzzling aspects of this group of viruses and is revealing genetic relationships among different virus serotypes.

The complete sequence of one virus (a US isolate of BTV-10) has been reported. The restriction sites of each BTV-10 gene as well as the predicted amino acid sequences of the encoded proteins have been included as an appendix, pp. 179–193.

In summary, our goal was to bring together basic and molecular aspects of virology and to develop a reference and source book for researchers and students at all levels. This could not have been accomplished without the contributions and support of the individual authors, to whom all credit is due.

Lastly, we are indebted to Professor Richard Compans, who is not only responsible for initiating this book, but has constantly provided constructive criticism whenever needed.

POLLY ROY
BARRY M. GORMAN

August 1989

List of Contents

B. M. GORMAN: The Bluetongue Viruses 1

H. HUISMANS and A. A. VAN DIJK: Bluetongue Virus Structural Components 21

P. ROY, J. J. A. MARSHALL, and T. J. FRENCH: Structure of the Bluetongue Virus Genome and Its Encoded Proteins 43

B. T. EATON, A. D. HYATT, and S. M. BROOKES: The Replication of Bluetongue Virus 89

I. M. PARSONSON: Pathology and Pathogenesis of Bluetongue Infections. 119

P. S. MELLOR: The Replication of Bluetongue Virus in *Culicoides* Vectors 143

J. L. STOTT and B. I. OSBURN: Immune Response to Bluetongue Virus Infection. 163

Subject Index 195

List of Contributors

(You will find the addresses at the beginning of the respective contribution)

BROOKES, S. M., p. 89
EATON, B. T., p. 89
FRENCH, T. J., p. 43
GORMAN, B. M., p. 1
HUISMANS, H., p. 21
HYATT, A. D., p. 89
MARSHALL, J. J. A., p. 43
MELLOR, P. S., p. 143
OSBURN, B. I., p. 163
PARSONSON, I. M., p. 119
ROY, P., p. 43
STOTT, J. L., p. 163
VAN DIJK, A. A., p. 21

The Bluetongue Viruses

B. M. Gorman

1 Bluetongue 1

2 The Emergent Disease 2

3 Bluetongue and Related Orbiviruses 4
3.1 The Bluetongue Virus Serogroup 5
3.2 The Epizootic Hemorrhagic Disease Virus Serogroup 8
3.3 The Umatilla Virus Serogroup 9
3.4 The Eubenangee Virus Serogroup 9
3.5 The Palyam Virus Serogroup 10

4 Diagnosis and Serology 11

5 Populations of Bluetongue Viruses 13

6 Concluding Remarks 14

References 15

1 Bluetongue

Bluetongue in sheep and cattle was first described in the late 18th century. Gutsche (1979) attributes the first description of "Tong-sikte" to a French zoologist, Francois de Vaillant, who travelled in the Cape of Good Hope between 1781 and 1784. Although clinical aspects of the disease were recorded by Hutcheon, the Chief Veterinary Officer of the Cape Colony, in his Annual Report for 1880, it was not until 1902 that "Malarial Catarrhal Fever" was first reported in the scientific literature (Hutcheon 1902).

Spreull (1905) wrote an account of the disease which he believed to be peculiar to South Africa. In a typical case description in sheep he indicated that the onset was marked by high fever lasting about 5–7 days. By 7–10 days distinctive lesions appeared in the mouth, and the tongue became severely affected and turned dark blue. He suggested that the word "malarial" was not applicable to the disease but that the common name "bluetongue" should be used. In his early reports Hutcheon had suggested that the agent was an insect-transmitted plasmodium. Robertson and Theiler (quoted by Spreull 1905)

United States Department of Agriculture, Agricultural Research Service, Arthropod-borne Animal Diseases Research Laboratory, P.O. Box 3965, University Station, Laramie, Wyoming, 82071-3965, USA

Current Topics in Microbiology and Immunology, Vol. 162

showed that the agent was filterable. SPREULL (1905) showed that the virus was transmitted to goats and to cattle and that the infection was inapparent.

Early attempts at vaccination included the use of serum from sheep which had recovered from the disease. THEILER (1908) immunized sheep by infection with a mild strain of virus which had been serially passaged in sheep. The original THEILER strain, now designated as serotype 4, was used for more than 40 years despite evidence that the vaccine was not safe and that the resultant immunity was not adequate. NEITZ (1948) was the first to recognize antigenically different types of bluetongue viruses (BTV) and to provide evidence for strain variation in virulence. In a series of cross-protection tests in sheep using ten strains isolated over a period of 40 years, NEITZ found that each strain produced solid immunity against reinfection, but variable protection against challenge with heterologous strains. A quadrivalent vaccine was introduced which included a strain now designated as serotype 12, two strains now designated as serotype 3, and the original THEILER strain (serotype 4). The serotyping of BTVs is based on serum-neutralization tests developed by HOWELL (1960, 1970). In South Africa the composition of bluetongue vaccines has been altered empirically to include new serotypes as they were detected (see HOWELL 1969). The present vaccine in use in South Africa is composed of a triad, each member containing five attenuated serotypes.

2 The Emergent Disease

Bluetongue was regarded as a disease of ruminants in Africa until 1943 when an outbreak occurred in Cyprus. According to GAMBLES (1949) there had been a number of outbreaks in Cyprus beginning in 1924, but in 1943–1944 a particularly virulent strain was responsible for about 2500 deaths in sheep. The mortality rate in flocks reached 70%. An antigenically distinct strain caused a further outbreak in 1946. GAMBLES commented that the same disease in a less severe form had been seen in Palestine in 1943 and in Turkey in 1944, 1946, and 1947. By 1951 bluetongue had been reported in Israel (KOMAROV and GOLDSMIT 1951); the strain was designated as serotype 4 (HOWELL 1969).

An apparently new disease entity of sheep known as "soremuzzle" was first recognized in Texas in 1948 (HARDY and PRICE 1952). The close resemblance to bluetongue was recognized, and BTV was isolated from cases of soremuzzle in sheep in California in 1952 (MCKERCHER et al. 1953). The virus was subsequently identified as serotype 10; viruses of that serotype had been previously identified in South Africa. BTV serotype 11 was isolated in New Mexico in 1955, serotype 17 in Wyoming in 1962, serotype 13 in Idaho and Florida in 1967, and serotype 2 in Florida in 1983. Serotype 17 viruses have been found only in the United States (USA).

In 1956 a major epizootic of bluetongue began in Portugal and extended into Spain (MANSO-RIBIERO et al. 1957). Within the first 4 months of the epizootic, a strain of serotype 10 virus caused 179 000 deaths in sheep with a mortality rate of 75% of affected sheep. A campaign of quarantine, slaughter, and compulsory annual vaccination produced a dramatic reduction in the incidence of disease such that it disappeared within 4 years, and no clinical cases have subsequently been recorded.

Bluetongue was reported in West Pakistan in 1958, in a flock of Rambouillet sheep previously vaccinated and imported from Utah, USA. HOWELL (1969) identified that virus as serotype 16 and reported that strains homologous in the serum-neutralization test were isolated in Israel in 1966. In describing an outbreak of bluetongue in goats and sheep in Maharashtra State, India, in 1961, SAPRE (1964) commented that the disease had caused havoc in sheep and goats in Pakistan in 1960.

The experience of bluetongue in the Iberian peninsula, the apparent spread of the disease through the Middle East and the Indian subcontinent, and the perception of bluetongue as an "emerging disease of animals" (HOWELL 1963) contributed to the restrictions placed on the movement of animals and importation of animal products. Perhaps in no country was threat of the introduction of BTV feared more than in Australia.

Before 1977 Australia was considered to be an area free of bluetongue disease. Strict quarantine regulations applied to livestock and animal products were designed to exclude exotic pathogens, including BTVs. Fears were often expressed of the serious economic losses which would be incurred if the viruses were introduced into Australia. A typical comment (BOWNE 1971) suggested that bluetongue had "explosive potential especially in countries like Australia where the sheep industry is of great economic importance and BT (bluetongue) could raise havoc with an extremely susceptible sheep population."

In a symposium on bluetongue held in Adelaide, Australia, in May 1974 many speakers referred to the potential dangers posed by BTVs. GEERING (1975) outlined the plans for control of bluetongue in an epizootic situation. Those plans were based on three lines of attack. These were "slaughter of ruminants in the infected area, disinsection of the infected area, [and] ...maintenance of a larger quarantine zone in which the standstill of ruminants ...was to be ...enforced." GEERING suggested that "a virulent strain of bluetongue could cause a mortality rate of up to 70% in the highly susceptible sheep population" and he proposed that it might "be necessary to blanket vaccinate a large portion of the Australian sheep population to lessen the serious effects of the disease." As part of that plan the Commonwealth Serum Laboratory held seed lots of most of the known serotypes of bluetongue which had been isolated in South Africa. If necessary, an appropriate vaccine could be made quickly in the event of an outbreak of bluetongue. In November 1977 it was announced that workers at the Commonwealth Scientific and Industrial Research Organization (CSIRO) Division of Animal Health had isolated a virus, from insects collected in the Northern Territory, which was indistinguishable from bluetongue virus (ST. GEORGE et al. 1978). Despite the fact the insects had been collected and the virus isolated more than 2 years before that announcement, and that there was no evidence of disease,

32 countries imposed bans on livestock imports, and control measures were introduced in parts of northern Australia (LEHANE 1981). Eight BTV serotypes have now been isolated in Australia. Three of these have so far been isolated only in Australia: type 20 in 1975, type 21 in 1979, and type 23 in 1982. Viruses of serotype 1 (1979), 15 (1982), 9 (1985), 3, and 16 (1986) have also been isolated in other countries (ST. GEORGE et al. 1980; GARD et al. 1987a, b).

3 Bluetongue and Related Orbiviruses

VERWOERD (1969) purified BTV and showed that the genome consisted of double-stranded RNA (dsRNA). He was the first to recognize the need to establish a new taxonomic group to include the known dsRNA viruses and those arthropod-borne viruses with physicochemical and morphological properties similar to those of BTV. Although not all the viruses he listed had been shown to contain dsRNA, VERWOERD (1970) proposed the name diplornaviruses for the new taxonomic group.

The International Committee on Taxonomy of Viruses decided that the bluetongue-like viruses would be defined by the genus *Orbivirus* within the family Reoviridae (FENNER 1976). The name *Orbivirus* had been proposed by BORDEN et al. (1971) to describe a number of arthropod-borne viruses that on the basis of morphological and physicochemical criteria formed a distinct group. The derivation from the Latin "orbis" ("ring" or "circle") was appropriate, since negatively stained virus particles when examined in the electron microscope have large doughnut-shaped capsomers.

VERWOERD et al. (1970), using polyacrylamide gel electrophoresis (PAGE), resolved the BTV genome into ten segments. The patterns of separation of the genome segments of a reovirus and a BTV were different, and the molecular weight of the BTV genome was estimated to be 12×10^6, compared with 15×10^6 for the reovirus (VERWOERD et al. 1970). Representative viruses from each of the recognized serogroups have been found to contain genomes similar to that of BTV (for a review, see GORMAN et al. 1983).

The described orbiviruses are differentiated into 12 serological groups (Table 1). No common generic antigen has been found, but viruses within each group share antigens detectable in complement-fixation (CF) tests, agar gel immunodiffusion tests, and fluorescent-antibody tests. The serotypes within each group are distinguishable by specific reactions in serum-neutralization tests.

The emphasis placed by different workers on the significance of cross-reactions in serological tests has created some confusion in the classification of orbiviruses. Within most of the serogroups, the serotypes cannot be distinguished by their reactions in the group-specific tests; in others the serotypes can be clustered within the serogroup by their cross-reactions in CF tests. The problem has been further complicated by reports of low-level cross-reactions among

Table 1. Orbivirus serological groups

Serogroup	No. serotypes
Bluetongue	24
Epizootic hemorrhagic disease	8
Umatilla	4
Eubenangee	2
Palyam	10
African horse-sickness	9
Equine encephalosis	7
Warrego	2
Wallal	3
Changuinola	12
Corriparta	6
Kemerovo[a]	23

[a] Kemerovo serogroup consists of four related but distinct groups (Brown et al. 1988b)

viruses that are normally considered to be members of distinct serogroups. There has been reluctance to include viruses of the Eubenangee and epizootic hemorrhagic disease (EHD) of deer serogroups in a BTV serogroup, despite many reports of serological relationships among viruses in these groups (Moore and Lee 1972; Moore 1974; St. George et al. 1978; Gorman and Taylor 1978; Della-Porta et al. 1985). BTVs are important pathogens of sheep, and it appears difficult to justify the inclusion of viruses, which are not known to cause disease in animals, in an extended BTV serogroup on the basis of low-level cross-reactions in certain serological tests.

The division of some serogroups into numbered serotypes and the formation of other serogroups with viruses of different names has also led to confusion in defining orbiviruses. Historically, the bluetongue and African horse-sickness groups were established by workers isolating viruses that were related to known serotypes but gave distinct reactions in cross-protection tests in animals or in serum-neutralization tests. Most of the other orbiviruses were isolated as by-products in programs to isolate and identify viruses as the causative agents of arthropod-transmitted disease. New viruses or new serotypes of a virus group are given names and are registered in the *International Catalogue of Arboviruses* (Karabatsos 1985). In this way, most of the orbivirus serogroups are named from the first member isolated and consist of viruses with different names. The use of names to describe serotypes in some serogroups and numbers to describe viruses of the African horse-sickness and bluetongue serogroups obscures the fact that the serogroups are composed of viruses related by the use of the same serological tests.

3.1 The Bluetongue Virus Serogroup

Following the demonstration of distinct types of BTVs in cross-protection tests in sheep (Neitz 1948) there were many attempts to develop serum-neutralization tests to study variation among BTV strains. Howell (1960, 1970) defined 16

Table 2. Prototype strains of bluetongue virus[a]

Serotype	Prototype strain	Isolation	
		Year	Country
1	Biggarsberg	1958	South Africa
2	Ermelo 22/59	1959	South Africa
3	Cyprus sample B	1943	Cyprus
4	Vaccine batch 603	1900	South Africa
5	Mossop	1953	South Africa
6	Strathene	1958	South Africa
7	Utrecht	1955	South Africa
8	Ermelo 89/59	1959	South Africa
9	University Farm	1942	South Africa
10	Ermelo 91/59	1959	South Africa
11	Nelspoort	1944	South Africa
12	Byenespoort	1941	South Africa
13	Mt. Currie	1959	South Africa
14	Ermelo 87/59	1959	South Africa
15	Onderstepoort 133/60	1960	South Africa
16	Hazara	1960	Pakistan
17	Wyoming 2790	1962	United States
18	—	—	South Africa
19	—	—	South Africa
20	CSIRO 19	1975	Australia
21	CSIRO 154	1979	Australia
22	—	1982	Australia
23	DDP 90	1982	Australia
24	—	—	South Africa

[a]Extended from HOWELL and VERWOERD (1971); GORMAN et al. (1983)

distinct antigenic groups of viruses, and subsequent application of similar techniques by many workers has led to the description of 24 different BTV serotypes (Table 2).

The significance of immunologically distinct serotypes in serum-neutralization tests is difficult to assess. There have been few attempts to correlate protection in animals with serotype, but there are indications that levels of neutralizing antibodies are a poor measure of protection (JEGGO 1986). The validity of the classification of BTVs into distinct serotypes may also be questioned. In early work, HOWELL (1960) used sheep convalescent sera and found some cross-neutralization between heterologous strains of BTV. These cross-reactions were virtually eliminated in subsequent tests by using hyperimmune guinea pig sera (HOWELL and VERWOERD 1971). THOMAS and TRAINER (1971) used convalescent and hyperimmune sera of calves to compare seven BTV strains in plaque-reduction tests. Convalescent sera cross-reacted with heterologous virus strains, but the cross-reactions were less than those observed using hyperimmune sera. BARBER and JOCHIM (1973) studied the serological characteristics of ten BTV strains isolated in the United States using plaque-reduction tests. On the basis of the reaction with hyperimmune sheep sera, the strains were classified into four serotypic groups, but cross-reactions among all ten strains were detected using hyperimmune rabbit sera.

The classification of BTVs into discrete groups seems improbable given the virtually continuous nature of biological variation. THOMAS et al. (1979) found extensive cross-reactions among BTV strains in North America in serum-neutralization tests and suggested that an antigenic continuum of virus strains existed instead of clearly defined antigenic groups. The serological comparison of the Australian strain CSIRO 19 with reference BTVs provides further evidence of the need to reassess the serum-neutralization tests used to identify BTV serotypes. Although the strain had been designated as a new serotype (serotype 20) by the World Reference Centre, The Veterinary Research Institute, Onderstepoort (VERWOERD et al. 1979), DELLA-PORTA et al. (1981) found the strain indistinguishable from serotype 4 using plaque-reduction, plaque-inhibition, or quantal microtiter neutralization tests. DELLA-PORTA et al. (1981) commented that the strain CSIRO 19 could be considered to be a subtype of serotype 4 and questioned its designation as a new serotype. Since animals inoculated with CSIRO 19 were protected against challenge with virulent serotype 4 or serotype 17 viruses, the practical value of a classification system based on in vitro reactions using carefully selected antisera must be questioned.

A wide variety of tests have been used in attempts to differentiate BTVs (for reviews, see VERWOERD et al. 1979; GORMAN et al. 1983). A passive hemagglutination (PHA) test for detection of BTV antibodies has been described (BLUE et al. 1974) in which partially purified virus is used to sensitize tannic acid-treated equine erythrocytes for testing; although some serotype specificity was found, the test has not been widely used.

JOCHIM and JONES (1980) developed a hemolysis in gel (HIG) test for BTVs. Like the CF test, the HIG test was group reactive, but BTVs and EHD viruses were differentiated in the test.

Hemagglutination which is specific for certain erythrocytes has been reported for BTVs (HUBSCHLE 1980; TOKUHISA et al. 1981a; van der WALT 1980; COWLEY and GORMAN 1987) and for EHD viruses (TOKUHISA et al. 1981b). In one study the BT8 strain from the USA (serotype 10) agglutinated sheep erythrocytes only, whereas strains of serotypes 3, 8 and 10 from South Africa agglutinated sheep, guinea pig, mouse, and chicken erythrocytes (HUBSCHLE 1980). An avirulent strain of serotype 10 from South Africa agglutinated sheep, goose, rabbit, and human erythrocytes (van der WALT 1980). A difference in the range of erythrocytes agglutinated by two BTVs isolated in Australia has also been reported. Bluetongue serotype 20 virus agglutinated sheep erythrocytes only, while serotype 21 virus agglutinated sheep, bovine, human, and goose erythrocytes (COWLEY and GORMAN 1987). The significance of the restricted range of erythrocytes agglutinated by some viruses and the extended range of others has not been investigated. Despite the serotype specificity in the hemagglutination and hemagglutination-inhibition test, it has not found wide use in serological tests for BTV infections.

MANNING and CHEN (1980) developed an enzyme-linked immunosorbent assay (ELISA) to detect antiviral IgG in sheep experimentally infected with four viruses of different serotype. The test was group reactive, as was a similarly based test described by HUBSCHLF et al. (1981).

3.2 The Epizootic Hemorrhagic Disease Virus Serogroup

Outbreaks of EHD have occurred in the USA since 1890. The New Jersey strain was isolated in 1955 and compared with the South Dakota strain isolated in 1956 (SHOPE et al. 1960). In cross-protection tests, deer that had recovered from infection with either strain resisted heterologous challenge. In neutralization tests, the viruses appeared to differ (METTLER et al. 1962), but effective comparison of the two strains was hampered by the lack of susceptible laboratory animals or cell culture for both viruses. The New Jersey strain was more lethal in deer than the South Dakota strain (SHOPE et al. 1960). The original isolate of the South Dakota strain has apparently been lost (R.E. SHOPE 1980, personal communication).

The virus has since isolated from white-tailed deer, mule deer, or antelope in the states of South Dakota, North Dakota, Michigan, Kentucky, Wyoming, Washington, North Carolina, Indiana, in Alberta, Canada, as well as from cattle and sheep in the state of Colorado (HOFF and TRAINER 1978; FOSTER et al. 1980; THOMPSON et al. 1988). The Alberta strain is designated as serotype 2 (BARBER and JOCHIM 1975).

Three strains of viruses from *Culicoides* spp. and one from *C. schultzei* colleted at Ibadan, Nigeria (LEE et al. 1974; LEE 1979) are probably distinct serotypes. The virus XBM/67 isolated in South Africa is referred to as a serotype (VERWOERD et al. 1979), but no detailed serological comparison with other EHD strains has been reported. Five distinct serotypes have been isolated in Australia (CAMPBELL and ST. GEORGE 1986). One of these (CSIRO 439) is identical to Ibaraki virus from Japan in serum-neutralization tests. Ibaraki virus is related to but distinct from the Alberta strain of serotype 2 (CAMPBELL et al. 1978) and produces a bluetongue-like illness in cattle (INABA 1975). In Table 3 serotypes 1 and 2 are designated according to convention. The virus strains from Australia, Nigeria, and South Africa are not numbered.

HUISMANS et al. (1979) found a 5%–10% homology between the genomes of the New Jersey strain of serotype 1 and a South African strain of bluetongue serotype 10. By cross-immune precipitation experiments common antigens were detected on VP7 and VP3 which are located in the nucleocapsids of the viruses (HUISMANS et al. 1979; HUISMANS and ERASMUS 1981).

BROWN et al. (1988a) examined the relationships among five EHD viruses on the basis of RNA–RNA hybridization. Nine of the ten genes of the viruses showed more than 74% homology. The most closely related were the New Jersey strain of serotype 1 and a strain from Nigeria (IbAr 22619), despite their geographical separation and the 12-year interval between virus isolations. The viruses cross-react in serum-neutralization tests (MOORE and LEE 1972; MOORE 1974). In hybridization tests EHD serogroup viruses were distantly related to BTV

Table 3. Prototype strains of epizootic hemorrhagic disease (EHD) virus

Serotype	Prototype	Isolation	
		Year	Country
1	New Jersey	1955	United States
2	Alberta	1962	Canada
	Ibaraki	1959	Japan
	CSIRO 439	1980	Australia
—	XBM/67	1967	South Africa
—	IbAr 22619	1967	Nigeria
—	IbAr 33853	1968	Nigeria
—	IbAr 49630	1970	Nigeria
—	CSIRO 157	1977	Australia
—	CSIRO 753	1981	Australia
—	CSIRO 775	1981	Australia
—	DPP 59	1982	Australia

serotype 10 from the USA, but one gene of serotype 10 consistently hybridized with gene 9 of the EHD viruses. The authors suggested that common antigenic determinants on the smallest nucleocapsid protein of the viruses (VP7) may be encoded on gene 9. In in vitro translation of RNA segments of EHD virus serotype 2, MECHAM and DEAN (1988) found that VP7 was encoded in gene 7 or 8. The discrepancy may reflect differences in technique for separating dsRNA genome segments.

Despite the low-level cross-reactions in some serological tests and cross-hybridization of at least one gene of EHD and BTVs, they probably represent two distinct populations of viruses. No evidence was found for gene reassortment in cell cultures infected simultaneously with viruses of both serogroups (GORMAN 1985; BROWN et al. 1988a) Gene reassortment has been demonstrated only between viruses of a serogroup and not between viruses in different serogroups of orbiviruses (reviewed in GORMAN 1983; GORMAN et al. 1983; KNUDSON and MONATH 1990).

3.3 The Umatilla Virus Serogroup

Viruses of the Umatilla serogroup have been isolated from species of *Culex* mosquitoes in the USA, Israel, and India. Umatilla, Llano Seco, and Netivot viruses are antigenically related to viruses in the bluetongue, EHD, and Eubenangee virus serogroups (TESH et al. 1986). The exact taxonomic status of the viruses is uncertain since cross-neutralization tests were inconclusive. A fourth virus, Minnal, isolated from *Culex vishuii* in India will be included in the serogroup (N. KARABATSOS, 1989 personal communication).

3.4 The Eubenangee Virus Serogroup

Eubenangee virus was isolated from a pool of 11 species of mosquitoes in northern Australia (DOHERTY et al. 1968). The virus reacted with antisera to EHD viruses and to Pata virus in CF tests, but reciprocal reactions were not detected (BORDEN et al. 1971). Pata virus had been isolated from *Aedes palpalis* in the Central African Republic. Despite low-level cross-reaction in these tests, Pata virus was recognized as a member of the Eubenangee serogroup, but neither virus was considered as a member of the EHD serogroup (KARABATSOS 1985). GONZALEZ and KNUDSON (1988) and BROWN et al. (1988a) compared Eubenangee, Pata, EHD, bluetongue, and certain other orbiviruses by RNA-RNA hybridization and by tests for gene reassortment between orbiviruses. They concluded that Pata virus was not related to Eubenangee or EHD viruses and recommended that the virus be placed in the ungrouped set of orbiviruses.

The third recognized serotype, Tilligerry virus, was isolated from *Anopheles annulipes* in New South Wales, Australia (GARD et al. 1973). In CF tests Tilligerry virus is more closely related to Eubenangee virus than to Pata virus (MARSHALL et al. 1980). Most of the isolations of Eubenangee group viruses have been made from pools of mosquitoes. In an extensive study involving isolation of viruses from arthropods collected in the Northern Territory of Australia, four isolations of Eubenangee-related viruses were made from *Culex annulirostris*, one from *Anopheles farauti*, and only one from the biting midge *Culicoides marksi* (STANDFAST et al. 1984). This contrasts with the isolations of BTVs and EHD viruses consistently from species of *Culicoides*. The serological relationships of these six isolates with the two serotypes of the Eubenangee serogroup have not been established, but GONZALEZ and KNUDSON (1988) showed that the four isolates from *C. annulirostris* were indistinguishable by RNA-RNA hybridization. All of the isolates cross-hybridized strongly in seven of the ten genes. DELLA-PORTA et al. (1979) reported that antiserum to the isolate from *A. farauti* neutralized infectivity of BTV serotype 1. There was no cross-hybridization of the RNA of any of the Eubenangee viruses with BTV RNA (GONZALEZ and KNUDSON 1988).

Viruses of the Eubenangee serogroup appear to be restricted to Australia. Despite reports of low-level cross-reactions in some serological tests between Eubenangee virus and BTVs (DELLA-PORTA et al. 1985), the viruses are distinct by RNA-RNA hybridization (BROWN et al. 1988a) and do not reassort genes on mixed infection in cell culture (TAYLOR 1984). The basis for shared antigens has not been established.

3.5 The Palyam Virus Serogroup

Viruses of the Palyam serogroup have been isolated in India, Australia, Japan, and Africa. Ten distinct serotypes have been defined (KNUDSON et al. 1984) which are closely related in RNA-RNA hybridization tests (BODKIN and KNUDSON 1986).

There have been consistent reports of cross-reactions in serological tests between viruses of the bluetongue and Palyam serogroups. MOORE (1974) reported cross-reactions in immunoprecipitin tests between Abadina, EHD, and bluetongue viruses. Sequential infection of cattle with four viruses of the Palyam serogroup led to the production of antibodies which reacted with BTV serotype 20 in immunoprecipitin tests (DELLA-PORTA et al. 1985). DELLA-PORTA et al. (1979) described an interesting situation in which serum from a cow experimentally infected with a Eubenangee virus neutralized BTV serotype 1 but not BTV serotype 20 or Ibaraki virus (EHD serogroup). The animal was subsequently challenged with a Palyam group virus and developed more cross-reactive antibodies that neutralized both BTV serotypes 1 and 20 and to a lesser extent, Ibaraki virus. The result illustrates the difficulty in interpreting serological tests on animals that may be exposed to infection with different orbiviruses.

4 Diagnosis and Serology

The diagnosis of bluetongue in sheep is uncomplicated for an experienced clinician, according to GIBBS (1983). However, the disease in cattle can be confused with infectious bovine rhinotracheitis, malignant cattarhal fever, and bovine virus diarrhea/mucosal disease. The diagnosis usually involves isolation of the virus or the detection of specific antibodies in the serum of a convalescent animal. There is no optimal procedure for isolating BTVs. Inoculation of susceptible sheep, embryonating chicken eggs, baby mice, and a variety of cells and cell lines in culture have been used. Some strains can be isolated directly in cell culture, but others can be isolated only in embryonating chicken eggs. No one system is best and JOCHIM (1985), after reviewing the procedures that have been used, concluded that there was more chance of isolating viruses if a number of systems were used.

Detection of antibodies to BTVs is based on group-reactive tests and the serotype-specific neutralization tests. A number of group-reactive tests including complement-fixation, fluorescent-antibody, and the agar gel immunodiffusion have been used. The CF test is often difficult to perform and a number of anticomplementary factors are encountered in its usual application. Addition of normal bovine serum to the test enhanced the level of BTV-specific antibody, and the modified CF test has been used extensively to certify animals free of antibodies to BTV (BOULANGER and FRANK 1975).

The agar gel immunodiffusion test for bluetongue was first described by KLONTZ et al. (1962) and has been used in a variety of modifications to detect antibodies to BTV. There are problems with the test in that it detects cross-reacting antibodies to other orbiviruses and it is relatively insensitive (DELLA-PORTA et al. 1985). The test is easy to perform and is inexpensive so that it is often preferred in surveying animals for BTV infection. In the interpretation of data obtained using the test, consideration has to be given to the period of persistence of antibodies. OSBURN et al. (1981) isolated viruses from 81 cattle, 35 of which were negative reactors in agar gel immunodiffusion tests, and from 122 sheep, 28 of which were negative in the tests.

Two types of ELISA have been used to detect BTV infections. In the indirect ELISA the viral antigen is coated on polystyrene plates and reacted with test serum or serum containing antibodies (MANNING and CHEN 1980; HUBSCHLE et al. 1981; POLI et al. 1982; LUNT et al. 1988; DROLET et al. 1988). A blocking ELISA has been described by ANDERSON (1984) in which immobilized antigen is reacted with a test serum and then with a group-specific murine monoclonal antibody. Antibody to BTV, if present in the test serum, blocks the antigen, preventing reaction with the monoclonal antibody in the last step of the test. Sera containing antibodies to EHD virus do not react in the test. Antibodies to cellular protein, which can complicate interpretation of immunodiffusion tests and indirect ELISA tests, do not interfere in the blocking ELISA. AFSHAR et al. (1987a) compared the indirect ELISA and a competitive ELISA for detection of antibodies to BTV. The competitive ELISA differed from the blocking ELISA only in that the monoclonal antibody was added to the reaction immediately after the addition of the test serum. They found the competitive ELISA more sensitive in the detection of antibodies than the indirect ELISA and as sensitive or more sensitive than the agar gel immunodiffusion test, the modified CF test, and the plaque neutralization test in detecting BTV antibodies. No reaction between the BTV antigen and an antiserum to EHD virus was seen in either test. Similar results were reported by LUNT et al. (1988) comparing a blocking ELISA using a monoclonal antibody to VP7 and an indirect ELISA for antibodies to BTV in experimental and field sera. The specificity of the blocking ELISA was absolute for BTV antibodies and showed no cross-reaction with antisera to EHD viruses. The blocking ELISA has also been used in a format in which the antigen was applied to nitrocellulose strips. Again, the ELISA was superior to the agar gel immunodiffusion test in detecting antibodies to BTV in experimental and field sera (ASHFAR et al. 1987b).

Detection of viral antigens in virus-infected ovine tissues using an indirect immunoperoxidase technique has been described (CHERRINGTON et al. 1985). The suggestion was made that the method could replace the fluorescent-antibody test, which generally lacks specificity, but the immunoperoxidase technique has not found wide application.

ADKINSON et al. (1988) studied the temporal development of humoral immune responses in sheep to natural BTV infection using Western immunoblotting. The procedure was superior to serum-neutralization and agar gel immunodiffusion tests in identifying past exposure to the virus. The immunoblotting procedure appeared to be group specific when assaying ruminant sera. Sera from BTV-infected sheep did not cross-react with EHD virus proteins and vice versa.

Cloned cDNA copies of RNA segments of a number of serotypes of BTV have been used to assess the relationships among viruses, but few have been used to detect BTV-specific RNA sequences in infected cells. A cDNA copy of segment 3 of serotype 17 hybridized to RNA from 19 serotypes of BTV but did not hybridize to the RNA of EHD virus serotype 1 (PURDY et al. 1984; ROY et al. 1985). The cDNA probe detected BTV-specific RNA in infected cells in culture and in red blood cells of experimentally infected sheep (ROY et al. 1985). The probe was also used in in situ cytohybridization to determine the tissue tropism and target cells for replication of BTV in the developing chick embryo. In contrast to the earlier hybridization results, the probe also detected RNA of EHD virus in infected embryos (Wang et al. 1988). A partial cDNA clone ($\sim 70\%$) of segment 7 of BTV serotype 17 was used in a dot-blot hybridization technique to detect BTV in cell culture (Squire et al. 1985). Despite the fact that the gene codes for the group-specific protein, VP7, the probe did not detect RNA of serotype 13 in infected cells.

5 Populations of Bluetongue Viruses

From RNA-RNA reassociation experiments it was apparent that geographical separation of BTVs had led to significant sequence divergence among them (Huismans and Howell 1973; Huismans and Bremer 1981; Gorman et al. 1981). These observations were extended using cDNA copies of certain genes in hybridization tests (Huismans and Cloete 1987) and in direct comparisons of the nucleotide sequences of some BTV genes. Gould (1988) defined three major groupings of the viruses—North American, Australian, and African—and suggested that there were other possible groupings based on serotype 15 isolated in Australia, serotype 16 isolated in West Pakistan, and serotype 3 from Cyprus. The observations are consistent with the evolution in isolation of distinct populations of BTVs and suggests the possibility that discrete gene pools evolve independently.

The concept of "gene pools" (see Adams 1979) of bluetongue and other orbiviruses implies populations of viruses able to exchange genetic information. Gene reassortment among orbiviruses has been demonstrated only between viruses of defined serogroups (reviewed in Gorman et al. 1983; Knudson and Monath 1990) except for the Kemerovo serogroup in which there are probably four reassortant groups (Brown et al. 1988b). Reassortant BTVs have been generated by experimental infections of cells in culture (Gorman et al. 1982; Kahlon et al. 1983) of sheep (Samal et al. 1987) cattle (Stott et al. 1987; Oberst et al. 1987), and of *Culicoides variipennis* (Samal et al. 1987). Despite these observations in experimental systems, the significance of the phenomenon in generating diversity in populations of BTVs has not been established. The only evidence for naturally occurring reassortants is based on comparative oligonucleotide fingerprinting of BTV isolates.

Sugiyama et al. (1981) compared oligonucleotide maps of genome segments of BTV serotypes 10 and 11, isolated in the USA, and concluded that one virus isolate was a natural reassortant between prototypes of serotypes 10 and 11. The oligonucleotide maps suggested that nine segments of the virus were derived from the serotype 10 virus and one (segment 3) from the serotype 11 virus. Segment 3 is highly conserved among BTVs (Gould 1987) and direct comparison of the nucleotide sequences of segments 3 of serotypes 17 and 10 isolated in the USA revealed 95.5% homology (Ghiasi et al. 1985). The interpretation of oligonucleotide maps of closely related prototype strains and a "naturally occurring" isolate to suggest reassortants in segment 3 should be viewed with caution.

Collisson and Roy (1983) compared the oligonucleotide maps of vaccine strains of serotype 10 used in the USA and concluded that one strain was a natural reassortant, and that the oligonucleotide map of segment 10 was more like the corresponding segment of a prototype 11 virus. The authors observed that "it appears very likely that reassortment between different BTV serotypes probably occurs continuously in nature although whether it takes place in the vectors or in the animal host (or both) is not known." The sequence of segment 10 of an Australian isolate of BTV-1 was more than 83% homologous at the nucleotide level with that of the US prototype strain of serotype 10 (Gould 1988), showing a greater conservation than that between segment 3 of the Australian virus and the US serotype 10 (Gould 1987). It is likely then that gene 10 of the US serotypes 10 and 11 is highly conserved. An interpretation of oligonucleotide maps of genome segments of closely related BTV isolates as indicative of natural reassortment is not convincing.

From the limited data available it is premature to infer that gene reassortment is a major determinant in the genetic structure of BTV populations. Caution seems justified when one considers the results of cross-hybridization of the genomes of BTVs isolated in Cyprus in 1971 and the original strain (serotype 4) isolated by Theiler in South Africa in 1900. Huismans and Howell (1973) detected nine hybrid molecules after hybridizing single-stranded RNA of the Cyprus isolate with the genome of Theiler's virus. Slight migrational differences in PAGE of the hybrid and native segments indicated that mutational change had occurred, but the genetic composition of the two viruses was remarkably conserved despite their geographical and temporal isolation. It is difficult to believe that such conservation could occur in freely reassorting BTV populations.

6 Concluding Remarks

The history of bluetongue in South Africa and the demonstration of its explosive potential in Spain and Portugal contributed to the belief that it would spread to the major sheep-farming areas of the world. The association of BTVs with cases of soremuzzle in sheep in the USA enhanced the proposition that bluetongue was one of the emerging diseases of animals (Howell 1963).

The discovery of BTVs in Australia was not the result of investigations of disease, but a by-product of a program designed to isolate viruses from the insect vectors of bovine ephemeral fever. Despite the fact that 8 of the 24 known BTV serotypes have now been isolated in Australia, there is still no evidence of clinical bluetongue in the country. That the viruses circulate "silently" suggests that they form part of a complex ecosystem of orbiviruses, insects, livestock, and other animals. Representative viruses of 7 of the 12 recognized serogroups of orbiviruses including EHD, Palyam, and Eubenangee have been isolated in Australia. Viruses of 3 of those 7 serogroups are apparently indigenous, as are 3 of the 8 serotypes of bluetongue and all but one of the serotypes of EHD, Palyam, and Eubenangee. The BTVs have probably evolved in Australia as part of a large complex of indigenous orbiviruses. Although periodic incursion of viruses is possible, the patterns of isolations of bluetongue and related orbiviruses suggest continuous circulation of viruses in defined regions.

A pattern similar to that seen in Australia appears to be emerging from a study of bluetongue in 11 countries in the Caribbean and Central America. Although there is no evidence of bluetongue disease in the region, more than 100 isolates have been made from healthy animals. Viruses of serotypes 1, 3, 6, and 12 have been identified, and these do not account for all of the potential animal-infecting serotypes which have been identified in serological tests (C. H. Barreto 1989, personal communication).

Although only a limited number of BTV genes have been sequenced, it was apparent to Gould (1988) that there were discrete BTV populations. Further analyses of the genetic structure of BTV populations will lead to a better understanding of the patterns of circulation of viruses. Analysis of archival strains should provide important information on the stability of the viral genomes and on the consequences of dispersal of viruses. The pattern of "emergence" of the disease from Africa can then be resolved.

Acknowledgements. This article was written during my term as Visiting Scientist at the Arthropod-borne Animal Diseases Research Laboratory. I thank Dr. T. E. Walton for organizing my appointment at the Laboratory. I am grateful to Mrs. Nancy Bozek for typing the manuscript and to Drs. Mecham, Luedke, Walton, Knudson, and Osburn for reviewing and commenting on it.

References

Adams MB, Coleman W (1979) From "gene fund" to "gene pool": on the evolution of evolutionary language. In: Studies in the history of biology, vol 3. Johns Hopkins University Press, Baltimore, pp 241–285

Adkinson MA, Stott JL, Osburn BI (1988) Identification of bluetongue virus protein-specific antibody responses in sheep by immunoblotting. Am J Vet Res 48: 1194–1198

Afshar A, Thomas FC, Wright PF, Shapiro JL, Shettigara PT, Anderson J (1987a) Comparison of competitive and indirect enzyme-linked immunosorbent assays for detection of bluetongue virus antibodies in serum and whole blood. J Clin Microbiol 25: 1705–1710

Afshar A, Thomas FC, Wright PF, Shapiro JL, Anderson J, Fulton RW (1987b) Blocking dot-ELISA using a monoclonal antibody for detection of antibodies to bluetongue virus in bovine and ovine sera. J Virol Methods 18: 271–280

Anderson J (1984) Use of monoclonal antibody in a blocking ELISA to detect group specific antibodies to bluetongue virus. J Immunol Methods 74: 139–149

Barber TL, Jochim MM (1973) Serological characterization of selected bluetongue virus strains from the United States. Proc Ann Meet US Anim Health Assoc 77: 352–359

Barber TL, Jochim MM (1975) Serotyping bluetongue and epizootic hemorrhagic disease virus strains. Proc 18th Ann Meet Am Assoc Vet Lab Diagn: 352–359
Blue JR, Dawe DL, Gatzek JB (1974) The use of passive hemagglutination for the detection of bluetongue viral antibodies. Am J Vet Res 35: 139–142
Bodkin DK, Knudson DL (1986) Genetic relatedness of Palyam serogroup viruses by RNA-RNA blot hybridization. J Gen Virol 67: 683–691
Borden EC, Shope RE, Murphy FA (1971) Physicochemical and morphological relationships of some arthropod-borne viruses to bluetongue virus—a new taxonomic group: physicochemical and serological studies. J Gen Virol 13: 261–271
Boulanger P, Frank JF (1975) Serological methods in the diagnosis of bluetongue. Aust Vet J 51: 85–189
Bowne JG (1971) Bluetongue disease. Adv Vet Sci Comp Med 15: 1–46
Brown SE, Gonzalez HA, Bodkin DK, Tesh RB, Knudson DL (1988a) Intra- and inter-serogroup genetic relatedness of orbiviruses. II. Blot hybridization and reassortment in vitro of epizootic hemorrhagic disease serogroup, bluetongue type 10, and Pata viruses. J Gen Virol 69: 135–147
Brown SE, Morrison HG, Buckley SM, Shope RE, Knudson DL (1988b) Genetic relatedness of the Kemerovo serogroup viruses: I. RNA-RNA blot hybridization and gene reassortment in vitro of the Kemerovo serocomplex. Acta Virol 32: 369–378
Campbell CH, St. George TD (1986) A preliminary report of a comparison of epizootic haemorrhagic disease viruses from Australia with others from North America, Japan and Nigeria. Aust Vet J 63: 233
Campbell CH, Barber TL, Jochim MM (1978) Antigenic relationship of Ibaraki, bluetongue and epizootic hemorrhagic disease viruses. Vet Microbiol 3: 15–22
Cherrington J, Ghalib H, Sawyer MM, Osburn BI (1985) Detection of viral antigens in bluetongue virus-infected ovine tissues, using the peroxidase-antiperoxidase technique. Am J Vet Res 46: 2356–2359
Collisson E, Roy P (1983) Analysis of the genomes of bluetongue virus serotype 10 vaccines and a recent BTV-10 isolate from Washington. Am J Vet Res 44: 235–237
Cowley JA, Gorman BM (1987) Genetic reassortants for identification of the genome segment coding for the bluetongue virus hemagglutinin. J Virol 61: 2304–2306
Della-Porta AJ, McPhee DA, Snowdon WA (1979) The serological relationships of orbiviruses. In St. George TD, French EL (eds) Arbovirus research in Australia, Proceedings, Second Symposium Brisbane, CSIRO-QIMR pp 64–71
Della-Porta AJ, McPhee DA, Snowdon WA (1979) The serological relationships of orbiviruses. In St George TD, French EL (eds) Arbovirus research in Australia, Proceedings, Second Symposium Brisbane, CSIRO-QIMR pp 64–71
Della-Porta AJ, Parsonson IM, McPhee DA (1985) Problems in the interpretation of diagnostic tests due to cross-reactions between orbiviruses and broad serological responses in animals. In: Barber TL, Jochim MM (eds) Bluetongue and related orbiviruses. Liss, New York, pp 445–453
Doherty RL, Standfast HA, Wetters EJ, Whitehead RH, Barrow GJ, Gorman BM (1968) Virus isolation and serological studies of arthropod-borne virus infections in a high rainfall area of north Queensland. Trans R Soc Trop Med Hyg 62: 862–867
Drolet BS, Mills KW, Belden EL, Mecham JO (1988) Development and application of an enzyme-linked immunosorbent assay (ELISA) for the detection of antibody to bluetongue virus. Proc USAHA 92: 113–122
Fenner F (1976) The classification and nomenclature of viruses. Intervirology 6: 1–12
Foster NM, Metcalf HE, Barber TL, Jones RH, Luedke AJ (1980) Bluetongue and epizootic hemorrhagic disease virus isolation from vertebrate and invertebrate hosts at a common geographic site. J Am Vet Med Assoc 176: 126–129
Gambles RM (1949) Bluetongue of sheep in Cyprus. J Comp Pathol 59: 176–190
Gard GP Marshall ID, Woodroofe GM (1973) Annually recurrent epidemic polyarthritis and Ross River virus activity in a coastal area of New South Wales. II. Mosquitoes, viruses and wildlife. Am J Trop Med Hyg 22: 551–560
Gard GP, Shorthose JE, Weir RP, Erasmus BJ (1987a) The isolation of a bluetongue serotype new to Australia. Aust Vet J 64: 87
Gard GP, Weir RP, Melville LF, Lunt RA (1987b) The isolation of bluetongue virus types 3 and 16 from northern Australia. Aust Vet J 64: 338
Geering WA (1975) Control of bluetongue in an epizootic situation: Australian plans. Aust Vet J 51: 220–224

Ghiasi H, Purdy MA, Roy P (1985) The complete sequence of bluetongue virus serotype 10 segment 3 and its predicted polypeptide compared with those of BTV serotype 17. Virus Res 3: 181–190
Gibbs EPJ (1983) Bluetongue disease. Agripractice 4: 31
Gonzalez HA, Knudson DL (1988) Intra- and inter-serogroup genetic relatedness of orbiviruses I. Blot hydridization of viruses of Australian serogroups. J Gen Virol 69: 125–134
Gorman BM (1983) On the evolution of orbiviruses. Intervirology 20: 169–180
Gorman BM (1985) Speciation in orbiviruses. In Barber TL, Jochim MM, Osburn BI (eds) Bluetongue and Related Orbiviruses, Prog Clin Biol Res 178 Alan Liss, New York, pp 275–278
Gorman BM, Taylor J (1978) The RNA genome of Tilligerry virus. Aust J Exp Biol Med Sci 56: 369–371
Gorman BM, Taylor J, Walker PJ, Davidson WL, Brown F (1981) Comparison of bluetongue type 20 with certain viruses of the bluetongue and Eubenangee serological groups of orbiviruses. J Gen Virol 57: 251–261
Gorman BM, Taylor J, Finnimore PM, Bryant JA, Sanger DV, Brown F (1982) A comparison of bluetongue viruses isolated in Australia with exotic bluetongue virus serotypes. In: St. George TD, Kay BH (eds) Arbovirus research in Australia: proceedings of the third symposium. CSIRO-QIMR, Brisbane, pp 101–109
Gorman BM, Taylor J, Walker PJ (1983) Orbiviruses. In: Joklik WK (ed) The reoviridae. Plenum, New York, pp 287–357
Gould AR (1987) The nucleotide sequence of bluetongue virus serotype 1 RNA 3 and a comparison with other geographic serotypes from Australia, South Africa, the United States of America and other orbivirus isolates. Virus Res 7: 169–183
Gould AR (1988) The use of recombinant DNA probes to group and type orbiviruses. A comparison of Australian and South African isolates. Arch Virol 99: 205–220
Gutsche T (1979) There was a man. Timmins, Cape Town, p 4
Hardy WT, Price DA (1952) Soremuzzle of sheep. J Am Vet Med Assoc 120: 23–25
Hoff GL, Trainer DO (1978) Bluetongue and epizootic hemorrhagic disease viruses: their relationship to wildlife species. Adv Vet Sci Comp Med 22: 111–132
Howell PG (1960) A preliminary antigenic classification of strains of bluetongue virus. Onderstepoort J Vet Res 28: 357–363
Howell PG (1963) Bluetongue. In: Emerging diseases of animals. FAO Agricultural Studies No. 61, Rome, pp 109–153
Howell PG (1969) The antigenic classification of strains of bluetongue virus, their significance and use in prophylactic immunization. D.V.Sc. thesis, University of Pretoria
Howell PG (1970) The antigenic classification and distribution of naturally occurring strains of bluetongue virus. J S Afr Vet Med Assoc 41: 215–223
Howell PG, Verwoerd DW (1971) Bluetongue virus. In: Hess WR, Howell PG, Verwoerd DW (eds) African swine fever virus, bluetongue virus. Springer, Berlin, Heidelberg, New York, pp 35–74 (Virology monographs, vol 9)
Hubschle OJB (1980) Bluetongue virus hemagglutination and its inhibition by specific sera. Arch Virol 64: 133–140
Hubschle OJB, Lorenz RJ, Matheka H-D (1981) Enzyme linked immunosorbent assay for detection of bluetongue virus antibodies. Am J Vet Res 42: 61–65
Huismans H, Howell PG (1973) Molecular hybridization studies on the relationships between different serotypes of bluetongue virus and on the difference between virulent and attenuated strains of the same serotype. Onderstepoort J Vet Res 40: 93–104
Huismans H, Bremer CW (1981) A comparison of an Australian bluetongue virus isolate (CSIRO 19) with other bluetongue virus serotypes by cross-hybridization and cross-immune precipitation. Onderstepoort J Vet Res 48: 59–67
Huismans H, Erasmus BJ (1981) Identification of the serotype-specific and group-specific antigens of bluetongue virus. Onderstepoort J Vet Res 48: 51–58
Huismans H, Cloete M (1987) A comparison of different cloned bluetongue virus genome segments as probes for the detection of virus-specified RNA. Virology 158: 373–380
Huismans H, Bremer CW, Barber TL (1979) The nucleic acid and proteins of epizootic hemorrhagic disease virus. Ondersterpoort J Vet Res 46: 95–104
Hutcheon D (1902) Malarial catarrhal fever of sheep. Vet Rec 14: 629–633
Inaba Y (1975) Ibaraki disease and its relationship to bluetongue. Aust Vet J 51: 178–184
Jeggo MH (1986) A review of the immune response to bluetongue virus. Rev Sci Tech Off Int Epiz 5: 357–362

Jochim MM (1985) An overview of diagnostics for bluetongue. In: Barber TL, Jochim MM (eds) Bluetongue and related orbiviruses. Liss, New York, pp 423–433
Jochim MM, Jones SC (1980) Evaluation of a hemolysis-in-gel test for detection and quantitation of antibodies to bluetongue virus. Am J Vet Res 41: 595–599
Kahlon J, Sugiyama K, Roy P (1983) Molecular basis of bluetongue virus neutralization. J Virol 48: 627–632
Karabatsos N (ed) (1985) International catalogue of arboviruses including certain other viruses of vertebrates. American Society of Tropical Medicine and Hygiene, San Antonio
Klontz GW, Svehag S-E, Gorham JR (1962) A study by the agar diffusion technique of precipitating antibody directed against blue tongue virus and its relation to homotypic neutralizing antibody. Arch Ges Virusforsch 12: 259–268
Knudson DL, Monath TP (1990) Orbiviruses. In: Fields BN, Knipe DM et al. (eds) Virology, 2nd ed. Raven, New York pp 1405–1433
Knudson DL, Tesh RB, Main AJ, St George TD, Digoutte JP (1984) Characterization of the Palyam serogroup viruses (Reoviridae: Orbivirus). Intervirology 22: 41–49
Komarov A, Goldsmit L (1951) A disease similar to bluetongue in cattle and sheep in Israel. Refuah Vet 8: 96–100
Lee VH (1979) Isolation of viruses from field populations of Culicoides [Diptera: Ceratopogonidae] in Nigeria. J Med Entomol 16: 76–79
Lee VH, Causey OR, Moore DL (1974) Bluetongue and related viruses in Ibadan, Nigeria: isolation and preliminary identification of viruses. Am J Vet Res 35: 1105–1108
Lehane L (1981) Bluetongue. Rur Res 112: 4
Lunt RA, White JR, Della-Porta AJ (1988) Studies with enzyme-linked immunosorbent assays for the serodiagnosis of bluetongue and epizootic haemorrhagic disease of deer. Vet Microbiol 16: 323–338
Manning JS, Chen MF (1980) Bluetongue virus: detection of antiviral immunoglobulin G by means of enzyme-linked immunosorbent assay. Curr Microbiol 4: 381–385
Manso-Ribeiro J, Rosa-Azevedo JA, Noronha F0, Braco-Forte-Junior MC, Grave-Periera C, Vasco-Fernandes M (1957) Fievre catarrhale du mouton (blue-tongue). Bull Off Int Epiz 48: 350–367
Marshall ID, Woodroofe GM, Gard GP (1980) Arboviruses of coastal south-eastern Australia. Aust J Exp Biol Med Sci 58: 91–102
McKercher DG, McGowan B, Howarth JA, Saito JK (1953) A preliminary report on the isolation and identification of the bluetongue virus from sheep in California. J. Am Vet Med Assoc 122: 300–301
Mecham JO, Dean VC (1988) Protein coding assignment for the genome of epizootic haemorrhagic disease virus. J Gen Virol 69: 1255–1262
Mettler NE, MacNamara LG, Shope RE (1962) The propagation of the virus of epizootic hemorrhagic disease of deer in newborn mice and HeLa cells. J Exp Med 116: 665–678
Moore DL (1974) Bluetongue and related viruses in Ibadan, Nigeria: serologic comparison of bluetongue epizootic hemorrhagic disease of deer and Abadina (Palyam) viral isolates. Am J Vet Res 35: 1109–1113
Moore DL, Lee VH (1972) Antigenic relationship between the virus of epizootic haemorrhagic disease of deer and bluetongue virus. Arch Ges Virusforsch 7: 282–284
Neitz WO (1948) Immunological studies on bluetongue in sheep. Onderstepoort J Vet Sci Anim Ind 23: 93–136
Oberst RD, Stott JL, Blanchard-Channell M, Osburn BI (1987) Genetic reassortment of bluetongue virus serotype 11 in the bovine. Vet Microbiol 15: 11–18
Osborn BI, McGowan B, Heron B, Loomis E, Bushnell R, Stott J, Utterback W (1981) Epizootiologic study of bluetongue: virologic and serologic results. Am J Vet Res 42: 884–887
Poli G, Stott JL, Liu YS, Manning JS (1982) Bluetongue virus: comparative evaluation of enzyme-linked immunosorbent assay, immunodiffusion and serum neutralization for detection of viral antibodies. J Clin Microbiol 15: 159–162
Purdy M, Petre J, Roy P (1984) Cloning the bluetongue virus L3 gene. J Virol 51: 754–759
Roy P, Ritter GD, Akashi H, Collisson E, Inaba Y (1985) A genetic probe for identifying bluetongue virus infections in vivo and in vitro. J Gen Virol 66: 1613–1619
Samal SK, Livingston CW, McConnell S, Ramig RF (1987) Analysis of mixed infection of sheep with bluetongue virus serotypes 10 and 17: evidence for genetic reassortment in the vertebrate host. J Virol 61: 1086–1091
Sapre SN (1964) An outbreak of bluetongue in goats and sheep in Maharashtra State, India. Vet Rev 15: 69–71

Shope RE, MacNamara LG, Mangold R (1960) A virus-induced epizootic hemorrhagic disease of Virginia white-tailed deer (Odocoileus virginianus). J Exp Med 111: 155–170
Spreull J (1905) Malarial catarrhal fever (bluetongue) of sheep in South Africa. J Comp Pathol Ther 18: 321–337
Squire KRE, Chuang RY, Chuang LF, Doi RH, Osburn BI (1985) Detecting bluetongue virus RNA in cell culture by dot hybridization with a cloned genetic probe. J Virol Methods 10: 59–68
Standfast HA, Dyce AL, St George TD, Muller MJ, Doherty RL, Carley JG, Filippich C (1984) Isolation of arboviruses from insects collected at Beatrice Hill, Northern Territory of Australia 1974–1976. Aust J Biol Sci 37: 351–366
St George TD, Standfast HA, Cybinski DH, Dyce AL, Muller MJ, Doherty RL, Carley JG, Filippich C, Frazier CL (1978) The isolation of a bluetongue virus from Culicoides collected in the Northern Territory of Australia. Aust Vet J 54: 153–154
St George TD, Cybinski DH, Della-Porta AJ, McPhee DA, Wark MC, Bainbridge MH (1980) The isolation of two bluetongue viruses from healthy cattle in Australia. Aust Vet J 56: 562–563
Stott JL, Oberst RD, Channell MB, Osburn BI (1987) Genome segment reassortment between two serotypes of bluetongue virus in a natural host. J Virol 61: 2670–2674
Sugiyama K, Bishop DHL, Roy P (1981) Analysis of the genomes of bluetongue viruses recovered in the United States. I. Oligonucleotide fingerprint studies that indicate the existence of naturally occurring reassortant BTV isolates. Virology 114: 210–217
Sugiyama K, Bishop DHL, Roy P (1982) Analysis of the genomes of bluetongue virus isolates recovered from different states of the United States and at different times. Am J Epidemiol 115: 332–347
Taylor J (1984) A study of genetic reassortment among Australian orbiviruses. M.Sc. thesis, University of Queensland
Tesh RB, Peleg J, Samina I, Margalit J, Bodkin DK, Shope RE, Knudson D (1986) Biological and antigenic characterization of Netivot virus, an unusual new orbivirus recovered from mosquitoes in Israel. Am J Trop Med Hyg 35: 418–428
Theiler A (1908) Inoculation of sheep against bluetongue and results in practice. Vet J 64: 600–607
Thomas FC, Trainer DO (1971) Bluetongue virus: some relationships among North American isolates and further comparisons with EHD virus. Can J Comp Med 35: 187–191
Thomas FC, Morse PM, Seawright GL (1979) Comparisons of some bluetongue virus isolates by plaque neutralization and relatedness tests. Arch Virol 62: 189–199
Thompson LH, Mecham JO, Holbrook FR (1988) Isolation and characterization of epizootic hemorrhagic disease virus from sheep and cattle in Colorado. Am J Vet Res 49: 1050–1052
Tokuhisa S, Inaba Y, Miura Y, Sato K (1981a) Salt-dependent hemagglutination with bluetongue virus. Arch Virol 70: 75–78
Tokuhisa S, Inaba Y, Miura Y, Sato K, Akashi H, Tsuda T, Shibata I (1981b) Hemagglutination of epizootic hemorrhagic disease virus. Arch Virol 69: 291–294
Van der Walt NT (1980) A haemagglutination and haemagglutination-inhibition test for bluetongue virus. Onderstepoort J Vet Res 47: 113–117
Verwoerd DW (1969) Purification and characterization of bluetongue virus. Virology 38: 203–212
Verwoerd DW (1970) Diplornaviruses: a newly recognized group of double-stranded RNA viruses. Prog Med Virol 12: 192–210
Verwoerd DW, Louw H, Oellermann RA (1970) Characterization of bluetongue virus ribonucleic acid. J Virol 5: 1–7
Verwoerd DW, Huismans H, Erasmus BJ (1979) Orbiviruses. In: Fraenkel-Conrat H, Wagner RR (eds) Comprehensive virology, vol 14. Plenum, New York, pp 285–345
Wang L, Kemp MC, Roy P, Collisson E (1988) Tissue tropism and target cells of bluetongue virus in the chicken embryo. J Virol 62: 887–893

Bluetongue Virus Structural Components

H. HUISMANS[1] and A. A. VAN DIJK[2]

1	Introduction	21
2	Viral Purification	22
3	Viral Morphology	22
4	Chemical Composition and Stability of the Virus	23
5	Viral Proteins	24
5.1	Structural Polypeptides	24
5.1.1	Outer Capsid Polypeptides	26
5.1.2	Core Polypeptides	29
5.2	Nonstructural Proteins	30
6	Viral Nucleic Acids	32
6.1	Double-Stranded RNA	32
6.1.1	Physicochemical Characteristics	32
6.1.2	Heterogeneity of the Genome and Genomic Probes	34
6.2	Messenger RNA	35
6.2.1	The Viral Transcriptase and mRNA Synthesis	35
6.2.2	mRNA Capping	36
7	Conclusion	36
	References	37

1 Introduction

The structural components of bluetongue virus (BTV), the prototype of the orbivirus genus, has been the subject of a number of reviews (VERWOERD et al. 1979; GORMAN and TAYLOR 1985; SPENCE et al. 1984). The main features can be summarized as follows: BTV is an icosahedral-shaped particle consisting of a segmented double-stranded RNA genome encapsidated in a double-layered protein coat. Removal of the outer protein layer activates a viral-associated RNA polymerase which transcribes the ten genome segments into 10 mRNAs which are in turn translated into at least seven structural and three nonstructural proteins. The characteristic features of these different structural components will be reviewed in this chapter. The key to the elucidation of the structural characteristics has been the ability to isolate and purify large amounts of BTV.

[1] Department of Genetics, University of Pretoria, Pretoria 0002, South Africa
[2] Biochemistry Section, Veterinary Research Institute, Onderstepoort 0110, South Africa

Current Topics in Microbiology and Immunology, Vol. 162

2 Viral Purification

The purification of BTV is hampered by difficulties in dissociating the virus from the cellular material to which it is very tightly bound as well as by the instability of purified virus particles. The first purification methods were based on homogenization of infected cells followed by fluorocarbon extraction, sucrose gradient centrifugation and fractionation on CsCl density gradients (VERWOERD 1969; VERWOERD et al. 1972; MARTIN and ZWEERINK 1972). Since the method is not equally satisfactory for the different serotypes and purified virus particles remain associated with varying amounts of either nonstructural viral proteins or small cellular proteins, alternative purification methods have been proposed. The cellular proteins are either absent or present in reduced amounts if the virus is dissociated from the cellular material by detergents such as Triton X-100 which do not disrupt the cell nucleus (HUISMANS et al. 1987b). MERTENS et al. (1987a) have also described an improved method to purify BTV and have recommended the use of sodium *N*-lauroyl sarcosinate to prevent the aggregation of virus particles.

3 Viral Morphology

Three morphologically distinct BTV particles namely virions, cores and subcore particles have been identified (Fig. 1). The different particles are obtained and the morphological features revealed by stepwise removal of some of the major structural proteins of the virus.

The negatively stained bluetongue virion is an icosahedral particle with a reported diameter of about 68–70 nm (VERWOERD et al. 1972; MARTIN and ZWEERINK 1972). The particles have a "fuzzy" appearance which clearly distinguishes BTV and other orbiviruses from reovirus and rotavirus which have

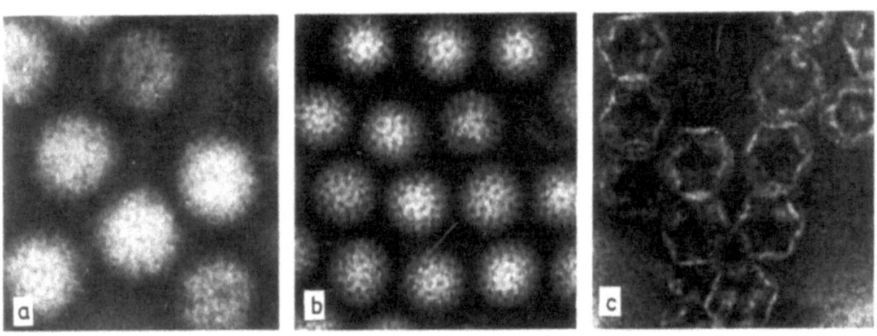

Fig. 1 a–c. Electron micrographs of **a** negatively stained bluetongue virus **b** core particles and **c** subcore particles. (Micrographs courtesy of Mr H.J. ELS, Electron Microscope Unit, Medical University of Southern Africa, 0204 Medunsa, South Africa)

a more structured and clearly defined outer capsid layer (PALMER et al. 1977). A sedimentation constant of 550S and a buoyant density in CsCl of 1.36 g/cm^2 has been reported for the bluetongue virion (MARTIN and ZWEERINK 1972; VAN DIJK and HUISMANS 1982). The density is similar to that reported for other orbiviruses such as African horse-sickness virus (AHSV) and epizootic haemorrhagic disease (EHD) virus (BREMER 1976; HUISMANS et al. 1979). A value of 1.38 g/cm^2 has, however, been reported by VERWOERD et al. (1972) and MERTENS et al. (1987a).

Bluetongue virions are converted to core particles by the removal of the outer capsid protein layer. The diameter of these core particles is in the order of 58 nm (MERTENS et al. 1987a) or 54 nm (ELS and VERWOERD 1969). Due to a tendency to aggregate, the core particles are difficult to purify on sucrose gradients. However, under the appropriate conditions they have a sedimentation constant of 470S. They are stable on CsCl gradients and have a density of 1.40 g/cm^2 (VAN DIJK and HUISMANS 1982; MERTENS et al. 1987a). The main morphological feature of the core particle is the presence of 32 distinct morphological units, or capsomeres with a circular configuration. These capsomeres are arranged in icosahedral symmetry with the triangulation number T-3 (ELS and VERWOERD 1969). They are tube-like, hollow structures about 8 nm long and 10–12 nm wide with an axial hole about 4 nm in diameter. This characteristic ring-like configuration of the capsomeres is the feature from which the genus name orbivirus is derived ("orbis" meaning ring or circle) (BORDEN et al. 1971).

In vivo almost all virions are uncoated to core particles immediately after infection. Later in the infection cycle a significant portion are further uncoated to subcore particles (HUISMANS et al. 1987d). These subcore particles are characterized by a skeleton-like appearance with a hexagonally shaped outline, resembling the subvirus particles of rotavirus described by ALMEIDA et al. (1979). The thin outer layer of the subcore particles has a side-to-side diameter of 40 nm and appears to be the base or scaffold on which the capsomeres are assembled. The capsomeres are presumed to be composed of major core protein VP7, which is the protein lost during the conversion from core to subcore particles. A method for the purification of subcore particles has been described (HUISMANS et al. 1987d), but it is evident that the particles are very unstable. In many respects the morphology of the BTV subcore particles bears a striking resemblance to the morphology and structure of double-stranded (ds)RNA phage Φ6 (ROMANTSCHUK et al. 1988). Both particles are composed of one major protein, three minor proteins, and dsRNA.

4 Chemical Composition and Stability of the Virus

The purified virion is composed of about 80% protein and 20% dsRNA (VERWOERD 1969). The base composition of the viral RNA yielded a C + G value of 42.4% (VERWOERD et al. 1979). There is no evidence that any single-stranded (ss) RNA is associated with the virion.

The stability of the virion has been reviewed by VERWOERD et al. (1979). A loss of infectivity is observed after the removal of one or more of the proteins in the outer capsid layer. Both monovalent and divalent cations can destabilize the outer capsid layer at ionic concentrations that are pH-dependent (HUISMANS et al. 1987b). The virions are most stable in the pH range of 8–9. Core particles are stable at a pH of as low as 5.0, but a further lowering of the pH results in a disruption of the particles. BTV is therefore not as stable over a wide pH range as reovirus (STANLEY 1967).

BTV virions are stable in lipid solvents (VERWOERD et al. 1979) and nonionic detergents such as Triton X-100 (HUISMANS et al. 1987b; MERTENS et al. 1987a). Nonpurified BTV particles in infected cells are strongly cell-associated and very stable at low temperatures over a long period. Purified virus particles are, however, unstable even at low temperatures and will rapidly lose infectivity. The problem is aggravated by a strong tendency of the particles to aggregate (MERTENS et al. 1987a).

5 Viral Proteins

The 10 dsRNA genome segments of BTV each code for the synthesis of at least one protein. The structural proteins are numbered VP1 to VP7 in order of decreasing size based on electrophoretic migration on polyacrylamide gels and

Table 1. Bluetongue virus proteins

Proteins	Coding[a] Segment	Location	Molecular[b] mass (dalton)	Total[c] structural proteins (%)	Estimated number/ virion[d]
VP1	1	Core	149 588	2.0	6
VP2	2	Outer capsid	111 023	22.7	97
VP3	3	Core	103 326	16.2	74
VP4	4	Core	76 433	0.9	5
VP5	6	Outer capsid	59 163	20.1	161
VP6	9	Core	35 750	2.8	37
VP7	7	Core	38 548	34.9	429
NS1	5	Infected cell	64 445	NA	NA
NS2	8	Infected cell	40 999	NA	NA
NS3	10	Infected cell	25 602	NA	NA

[a] Assignments as suggested by PEDLEY et al. 1988 based on reports by MERTENS et al. (1984) as well as VAN DIJK and HUISMANS (1988). Segments are numbered in order of decreasing size which agrees with their order of migration in 1% agarose gels (PEDLEY et al. 1988)
[b] Calculated from nucleotide sequences obtained for cloned genome segments (ROY 1989)
[c] From VERWOERD et al. (1972)
[d] Calculated as described by VERWOERD et al. (1972) but using the molecular mass values of the cloned genome segments determined by ROY (1989)
NA, not applicable

the nonstructural proteins are designated NS1, NS2, and NS3. However, in the case of some BTV serotypes the order of migration of VP2 and VP3 is reversed (GRUBMAN et al. 1983). It has also been reported that the order can be reversed if different polyacrylamide cross-linkers are used (MECHAM et al. 1986). The coding assignments, as determined by in vitro translation of the positive strand of each of the ten dsRNA genome segments of the virus, have been reported for BTV-17, BTV-1, and BTV-10 by GRUBMAN et al. (1983), MERTENS et al. (1984) and VAN DIJK and HUISMANS (1988), respectively. These results are summarized in Table 1. In view of the fact that different electrophoretic conditions can affect and even reverse the order of migration of the genome segments (KOWALIK and LI 1987), the genome segments of BTV are numbered 1 to 10 in order of decreasing size on agarose gels, as suggested by PEDLEY et al. (1988). The characterization of each of the seven structural and three nonstructural proteins has been a major part of the continuing research on BTV.

5.1 Structural Polypeptides

There appears to be general consensus that the fully infectious BTV virion contains four major proteins (VP2, VP3, VP5 and VP7) and three minor proteins (VP1, VP4 and VP6) (VERWOERD et al. 1972; MARTIN and ZWEERINK 1972). Of these, major proteins VP2 and VP5 form the diffuse outer capsid layer of the virus.

There have been various reports of the association of purified virus with either cellular or nonstructural proteins. These include the small nonstructural proteins NS3 and NS3a (MERTENS et al. 1984) as well as NS2 (MERTENS et al. 1987a). MECHAM et al. (1986) on the other hand found that some of the BTV serotypes are still associated with nonstructural protein NS1 after purification. More compelling evidence has been provided by EATON et al. (1988), who used an immunogold labelling procedure to show that both virus and core particles contain NS1. Judged by the position of the gold-labelled probes, NS1 may protrude from the surface of the core particles. The immunogold labelling results were confirmed by analysis of purified virus and core particles. The authors suggested that the Triton X-100 purification method that they used could be responsible for the discrepancy between their results and those obtained by VERWOERD et al. (1972). However, MERTENS et al. (1987a) and HUISMANS et al. (1987b) have both used Triton X-100 purification methods but failed to detect similar amounts of NS1. It remains to be clarified whether incorporation of NS1 into the capsid is a functional necessity or a coincidental event.

The molecular mass of the various BTV proteins was initially determined by their migration on SDS-containing polyacrylamide gels (VERWOERD et al. 1972; DE VILLIERS 1974). These results are flawed to some extent by the fact that the relative order of migration can be influenced by electrophoretic conditions such as the buffer system and acrylamide concentration (HUISMANS and BREMER 1981;

MERTENS et al. 1984). More recently the exact size of the BTV-10 (US) polypeptides have been determined after sequencing the corresponding cloned genome segments. These results are discussed in Chap. 3. In Table 1 the relative molar amount of the seven structural proteins are indicated. These values were first determined by VERWOERD et al. (1972), but since a much better estimate of the size of the different proteins is now available the molar ratios were recalculated using the size of the proteins shown in Table 1. The major capsid proteins constitute about 93% of the protein content of the virus. Proteins VP1 and VP4 are present in the smallest relative amount of about 5–6 copies/virion each, whereas VP7 is present in the largest number of copies.

There is no evidence that any of the capsid proteins are modified either by phosphorylation or glycosylation. One possible exception is protein VP6, a component of the core particle, which is often resolved into two bands of very similar electrophoretic mobility on PAGE gels (HUISMANS and BREMER 1981; MERTENS et al. 1984). The nature of this doublet is unknown, but it is possible that one is a modified form of the other. It is not observed in all the BTV serotypes and has for example never been observed in the case of BTV-10.

Most of the proteins are synthesized throughout the infection cycle with a relative frequency that is not significantly different from that in which the corresponding mRNA species are synthesized (HUISMANS 1979; HUISMANS et al. 1987c). The first virus-specific polypeptides are detected 2–4 h after infection and there is a rapid increase in the rate of synthesis until about 11–13 h p.i. after which the rate remains more or less constant until at least 24 h after infection.

With respect to the function and individual characteristics of the seven structural proteins we will distinguish between outer capsid and core polypeptides.

5.1.1 Outer Capsid Polypeptides

The outer capsid layer of BTV consists of two major polypeptides, VP2 and VP5, which together constitute approximately 40% of the total protein content of the virus. Little is known about the interaction between these two proteins, their topography and degree of exposure on the surface of the virus particle. However, the fact that VP2 can be dissociated from the virion without removal of VP5 (VERWOERD et al. 1972; HUISMANS et al. 1987b) indicates that VP5 is more closely associated with the core particle than VP2. If VP5 has to recognize or bind both VP2 and one of the core proteins, it would place a higher degree of constraint on the variability of VP5 than on VP2. The variability of these two proteins has been the subject of a number of investigations

It was demonstrated by PAGE that VP2 and VP5 show the largest variation in size among the structural proteins of the different BTV serotypes (DE VILLIERS 1974; MECHAM et al. 1986). Peptide mapping has furthermore indicated that VP2 is unique for each of the US serotypes, whereas VP5 showed an intermediate level of conservation (MECHAM et al. 1986). These results have been substantiated by data on the nucleic acid homology of the genes coding for the outer capsid polypeptides. These results, which are reviewed in Chap. 3, include mRNA/dsRNA hybridization between BTV serotypes (HUISMANS and HOWELL 1973), hybridization of labelled dsRNA, or small dsRNA fragments of one serotype to Northern blots of dsRNA of another serotype (SQUIRE et al. 1986; KOWALIK and LI 1987; MERTENS et al. 1987b) and DNA/RNA hybridization of cloned genomic probes with dot spots or Northern blots of dsRNA (HUISMANS and CLOETE 1987; HUISMANS et al. 1987a; GOULD 1988a; UNGER et al. 1988a, b). Comparisons of amino acid sequences of VP2s and VP5s from several BTV serotypes have also been reported (GHIASI et al. 1985; PURDY et al. 1986; FUKUSHO et al. 1987; GOULD 1988a; GOULD and PRITCHARD 1988 and WADE-EVANS et al. 1988). The high degree of conservation of VP5 polypeptides in comparison with VP2 reflects a high degree of restraint on the structural variability of VP5.

The variability of the major outer capsid polypeptides of BTV reflects the role of these proteins in the induction of serotype-specific neutralizing antibodies. While most of the evidence would indicate that VP2 is the main determinant of serotype-specificity, a contributory role of VP5 can by no means be excluded.

The first evidence for the role of VP2 in determining serotype specificity was obtained by demonstrating that VP2 immune precipitation was serotype specific (HUISMANS and ERASMUS 1981). Similar results were obtained with the five BTV serotypes isolated in the USA (MECHAM et al. 1986).

More direct evidence that VP2 is involved in the induction of serotype-specific antibodies was obtained by demonstrating that VP2 and a mixture of VP2 and VP5 can elicit neutralizing antibodies in rabbits and sheep (HUISMANS et al. 1987b). The immune response obtained after injection of sheep with 100 μg purified VP2 was sufficient to protect them against challenge with virulent virus of the same serotype.

VP2 purified from gels after SDS–PAGE failed to induce neutralizing antibodies. However, neutralizing antibodies were obtained by VP2 synthesized in cells infected with baculovirus recombinants containing the cloned segment 2 gene of BTV-10 (US) (INUMARU and ROY 1987). These results indicate that the neutralization epitope could be conformation dependent. The importance of VP2 in the induction of a neutralization-specific immune response was further confirmed by the isolation of VP2-specific monoclonal antibodies which can neutralize the virus (APPLETON and LETCHWORTH 1983) and provide passive protection against challenge with a homologous virulent BTV strain (LETCHWORTH and APPLETON 1983).

The regions on VP2 that are involved in determining serotype specificity have not as yet been identified. A comparison of the amino acid sequence of VP2 from two BTV serotypes failed to locate the antigenic site(s) involved in the serotype-specific variation (FUKUSHO et al. 1987). GOULD et al. (1988) have isolated a number of escape mutants using a neutralizing monoclonal antibody against BTV-1 (Aus). When compared with the wild-type virus the mutants each revealed a single nucleotide substitution in the segment 2 gene which resulted in a single amino acid change in VP2. The changes were all located between amino acid positions 328 and 335 in VP2. This variable region is flanked by sequences which contain a mixture of conserved and variable residues which in turn are flanked by two regions of highly conserved sequences. The conserved regions could be very important in ensuring that the overall three-dimensional conformation of the variable serotype-specific region is maintained. The region containing the neutralization-specific mutations (GOULD et al. 1988) is outside the variable VP2 regions identified by GHIASI et al. (1987). It is, however, highly likely that more than one serotype-specific epitope is involved. For example, two VP2-specific monoclonal antibodies, one of which neutralizes BTV-1 (Aus) but not BTV-1 (SA) and another one which neutralizes both, have been reported (WHITE and EATON as reported by GOULD et al. 1988).

The possibility that VP5 contributes to the neutralization-specific immune response has not been fully explored. KAHLON et al. (1983) have used reassortants to demonstrate that serotype specificity cosegregates with the VP2 but not with the VP5 genome segments. In a study of reassortant viruses of BTV types 20 and 21, it was shown that reassortants which possessed both outer capsid proteins VP2 and VP5 were neutralized specifically by homologous antiserum. However, reassortants with VP2 of BTV-20 and VP5 of BTV-21 had intermediate characteristics (COWLEY and GORMAN 1989). In similarly designed studies, MERTENS et al. (1989) isolated a reassortant containing VP5 from BTV-3 and VP2 from BTV-10 which cross-neutralized with both parental virus strains (BTV-3 and BTV-10), indicating a possible contributory role for VP5 in the induction of neutralizing antibodies. There is, however, as yet no evidence that VP5 by itself can induce neutralizing antibodies.

Other functions of the virus that are generally thought to be associated with proteins in the outer capsid layer are virulence and cell adsorption. RNA/RNA hybridization of a number of virulent and avirulent strains of homologous serotypes (HUISMANS and HOWELL 1973) indicated the occurrence of very small differences in those genome segments coding for the outer capsid polypeptides. WALDVOGEL et al. (1986) have isolated two strains of BTV-11 that differed significantly in their virulence for newborn mice. The two strains were distinct electropherotypes with the most obvious difference in the electrophoretic mobility of genome segment 5.

There is some evidence that VP2 is involved in cellular attachment. A loss of infectivity is associated with the removal of VP2 (VERWOERD et al. 1972) and such particles can also no longer bind to cells (HUISMANS et al. 1983). Intermediate subviral particles of BTV that contained VP2 chymotrypsin cleavage products were, however, fully infectious (MERTENS et al. 1987a). Although cleavage of VP2 did not affect cell attachment, it did result in a loss of haemagglutination activity. This function is generally also associated with VP2 since haemagglutination inhibiting antibodies to BTV are type-specific (TOKUHISA et al. 1981; HÜBSCHLE 1980; VAN DER WALT 1980). The results suggest that the sites for haemagglutination activity and cell attachment are not necessarily the same. COWLEY and

GORMAN (1987) have used reassortants to locate the haemagglutination activity on VP2.

5.1.2 Core Polypeptides

The BTV core is composed of two major polypeptides (VP3 and VP7) and three minor polypeptides (VP1, VP4 and VP6). VP7 predominates and comprises approximately one-third of the total protein content of the virion (Table 1).

VP7 is the main component of the capsomeres on the surface of the BTV core particle (HUISMANS et al. 1987d). There is also some evidence that the protein is not completely shielded by the outer capsid and can be recognized in the double-layered virus particle by VP7-specific antibodies (HYATT and EATON 1988). VP3 on the other hand is not recognized by antibodies, suggesting an inner location. VP3 is the only major structural protein in the BTV subcore particles and it has been proposed that it forms the protein scaffold on which the capsomeres are arranged (HUISMANS et al. 1987d).

VP7 has been identified as the soluble group-specific antigen (HUISMANS and ERASMUS 1981; GUMM and NEWMAN 1982; HÜBSCHLE and YANG 1983). The protein has been purified by chromatofocusing (GUMM and NEWMAN 1982) and has been characterized by peptide mapping (WHISTLER and NEWMAN 1986). Tryptic maps of VP7 from seven different BTV serotypes indicated that VP7 from some serotypes was only distantly related. A similar pattern was also reflected by the nucleic acid homology of cognate genome segments that encode VP7. In contrast to those genome segments that code for the other core proteins which were found to be highly conserved (more than 90% nucleic acid homology), the variation in the VP7 genome was much larger (HUISMANS and CLOETE 1987; HUISMANS et al. 1987a; KOWALIK and LI 1987; RITTER and ROY 1988). In a number of BTV serotypes the VP7 genome homology was high (more than 90%), whereas between others the homology was much less. Since VP7 is the most exposed of the core polypeptides it might well be subjected to immunological pressure which could explain the variation in VP7 that is observed.

Little is known about the minor capsid polypeptides VP1, VP4 and VP6. The nucleotide sequence of BTV-10 (US) segment 1 which encodes VP1 has recently been determined (ROY et al. 1988). VP1 was shown to be a 149 588 dalton polypeptide with extensive homology to a vaccinia virus DNA-dependent RNA polymerase subunit, as well as several other prokaryotic and eukaryotic RNA polymerases. VP1 and VP4 are present in very small amounts of about 5–6 copies per virus particle and a cooperative enzymatic function of the two proteins in RNA transcription and/or RNA replication is a distinct possibility. Some of these enzymatic activities will be discussed in more detail in Sects. 6.2.1 and 6.2.2.

Nothing is known about protein VP6 except that it migrates as a doublet in several of the BTV serotypes. It is also surprising that the molecular weight of VP6 is smaller than that of VP7 (Table 1), although it migrates in a position above VP7. The reason for this is not clear.

5.2 Nonstructural Proteins

At least three different nonstructural proteins have been identified in BTV-infected cells. The two major nonstructural proteins, NS1 (P5A) and NS2 (P6A), were first identified by HUISMANS (1979). A third minor nonstructural protein, NS3, was first detected by GORMAN et al. (1981) in BTV-infected cells.

The synthesis of nonstructural proteins NS1 and NS2 in BTV-infected cells coincides with that of the structural proteins (HUISMANS 1979). Protein NS1 is synthesized in large amounts, comprising about 25% of the virus-specified proteins in the infected cells. However, very little NS1 is detected in the soluble fraction of infected cells because it is very rapidly converted to high molecular weight tubular structures with a sedimentation coefficient of about 300S–500S (HUISMANS 1979; HUISMANS and ELS 1979). The presence of these tubules is a characteristic feature of BTV-infected cells (LECATSAS 1968). Tubules are also found in cells infected with other orbiviruses such as EHDV and AHSV (TSAI and KARSTAD 1970; OELLERMAN et al. 1970). Negatively stained BTV tubules are approximately 68 nm in diameter and are characterized by a surface fine structure with a striking 9 nm linear periodicity along the length of the tubules, giving them a ladder-like appearance. The EHDV tubules are 54 nm in diameter whereas the AHSV tubules are much smaller with a diameter of 18 nm and no clearly defined surface structure (HUISMANS and ELS 1979).

The tubules are distinct from the microtubules in uninfected cells. Colchicine, which inhibits the polymerization of tubulin into microtubules (WILSON and MESA 1973), had no effect on virus tubule formation in BTV- and AHSV-infected cells (HUISMANS and ELS 1979; EATON et al. 1987). Polymerization of NS1 into tubules occurs throughout the infection cycle and can be demonstrated as early as 2–4 h p.i., long before the appearance of infectious progeny virus particles. Recently URAKAWA and ROY (1988) have expressed the gene that codes for NS1 by means of a baculovirus recombinant. The protein was synthesized in large amounts and formed numerous tubules in *Spodoptera frugiperda* cells.

EATON et al. (1988) have investigated the localization of NS1 in BTV-infected cells, using NS1-specific monoclonal antibodies. These results have been reviewed by EATON and HYATT (1989) and are discussed in a later chapter. A group of NS1-specific monoclonal antibodies was identified which reacted with virus particles in infected cells. These particles were either leaving or in close proximity to the viral inclusion bodies. It seems possible that some NS1 remains associated with highly purified virus or core particles.

The gene coding for NS1 (segment 5) is transcribed more frequently than any of the other genome segments (HUISMANS and VERWOERD 1973). This could explain the high relative amounts in which NS1 is synthesized. The fact that segment 5 is more frequently expressed is not unique to BTV but has also been observed in the case of other orbiviruses such as EHDV (HUISMANS et al. 1979). It is unknown why such large amounts of NS1 are required in the infected cells. An attractive hypothesis presented by EATON et al. (1988) is that the tubular structures, which appear to condense from NS1-rich fibrillar material, are the repository of NS1 that has been utilized in a prior stage of virus morphogenesis. If the tubules themselves are not the active participants in virus morphogenesis there could be a necessity for a continued large supply of fresh, soluble NS1 throughout the infection cycle.

The gene coding for NS1 is highly conserved in the BTV serogroup (Huismans and Cloete 1987). It has been sequenced by Lee and Roy (1987) and was shown to code for a 553 amino acid protein that does not appear to have any similarity to known microtubular proteins. In addition to NS1, Mertens et al. (1984) have observed another protein called NS1a which is also coded for by genome segment 5 (Pedley et al. 1988). The relationship between NS1a and NS1 is not clear.

The function of NS2, the other major nonstructural protein, is also not clear. It has some features in common with protein σ-NS of reovirus. The two proteins are of approximately the same size and both have affinity for ssRNA (Huismans and Joklik 1976; Huismans et al. 1987c). Such ssRNA-binding nonstructural proteins appear to be common to all members of the Reoviridae family and have also been found in the case of rotavirus (Boyle and Holmes 1986). A distinctive feature of the BTV NS2 is that it is phosphorylated (Huismans et al. 1987c). NS2 purified by affinity chromatography can also be phosphorylated in vitro without addition of exogenously added phosphokinase, suggesting that the kinase that is responsible for phosphorylation remains associated with NS2 during purification (Huismans et al. 1987c)

NS2 was detected in both the soluble and particulate fraction of infected cells. Particulate NS2 can be solubilized by a high-salt treatment and can then be purified using affinity chromatography (Huismans et al. 1987c). It was also found that mixtures of BTV mRNA and soluble NS2 form a complex with an estimated S value of about 22 on sucrose gradients. The S value was independent of the mRNA/NS2 ratio. The exact stochiometric ratio or molar amounts of the macromolecules in these complexes have not as yet been determined. An attractive hypothesis for the function of NS2 is that it acts in the selection and condensation of the ten mRNA species during virus morphogenesis. Eaton et al. (1988) refer to an unpublished observation by A.D. Hyatt that NS2 is associated with the viral inclusion bodies in the cytoskeleton of BTV-infected cells. Mertens et al. (1987a) have reported that a small amount of NS2 remains associated with highly purified virus. The gene coding for NS2 has been cloned and sequenced (Hall et al. 1989). It codes for a protein of 357 amino acids.

The third nonstructural protein, NS3, is encoded by the smallest of the BTV genome segments (Mertens et al. 1984; Van Dijk and Huismans 1988). The in vitro translation product consists of two proteins NS3 and NS3A with a molecular weight of 28 000 and 25 000 daltons respectively. The two proteins have almost identical peptide maps and were also found to be synthesized in very small amounts in BTV-infected cells (Van Dijk and Huismans 1988). Whether NS3 and NS3A are functional equivalents or complement one another is not known. The NS3-coding genome segment of BTV-10 (US) and BTV-1 (Aus) have been cloned and sequenced (LEE and ROY 1986; GOULD 1988b) indicating two in-phase, overlapping open reading frames.

6 Viral Nucleic Acids

BTV nucleic acids are comprised of dsRNA and single-stranded mRNA. The mRNAs are transcribed from the dsRNA template and each of the ten mRNA species is an exact copy of the positive strand of one of the dsRNA segments (HUISMANS and VERWOERD 1973). There are no small oligonucleotides associated with BTV as is the case with reovirus (BELLAMY et al. 1967; SHATKIN and SIPE 1968). In this section we will summarize the characteristics of the various viral RNAs, the synthesis of mRNA, and the use of genomic probes for the detection of viral RNA.

6.1 Double-Stranded RNA

6.1.1 Physicochemical Characteristics

Rapid progress has been made in the characterization of the BTV dsRNA segments in the past few years. All the ten genome segments of BTV-10 (US) have been cloned and sequenced and these results will be summarized in a following chapter. Important common features of all the dsRNA segments are that the 5'- and 3'-noncoding regions are relatively short, varying between 8 and 34 bp (ROY 1989). A sequence of six nucleotides at the 3'- and 5'-ends of all BTV dsRNA segments is conserved (KIUCHI et al. 1983; RAO et al. 1983; MERTENS and SANGAR 1985).

There are considerable differences in the PAGE migration profiles of the genome segments of the different serotypes of BTV (Fig. 2a). These migration patterns are not characteristic for different isolates of the same serotype (OBERST et al. 1987; MERTENS et al. 1987b) (Fig. 2a) and can therefore not be used for conclusive serotype classification (GORMAN and TAYLOR 1985). The variation in electrophoretic migration on PAGE gels is furthermore not an accurate reflection of the size of the genome segments (PEDLEY et al. 1988). It has been shown that the relative order of migration of the genome segments is affected by and can even be reversed by variation in polyacrylamide concentrations (MERTENS and SANGAR 1985; KOWALIK and LI 1987). The differences can be ascribed to variations in base composition or to secondary structure and are less distinct at low polyacrylamide concentrations.

On agarose gels the dsRNA genome segments are separated according to size and no differences in the dsRNA profiles of the different BTV serotypes have been reported (Fig. 2b) (SQUIRE et al. 1983; KOWALIK and LI 1987; PEDLEY et al. 1988).

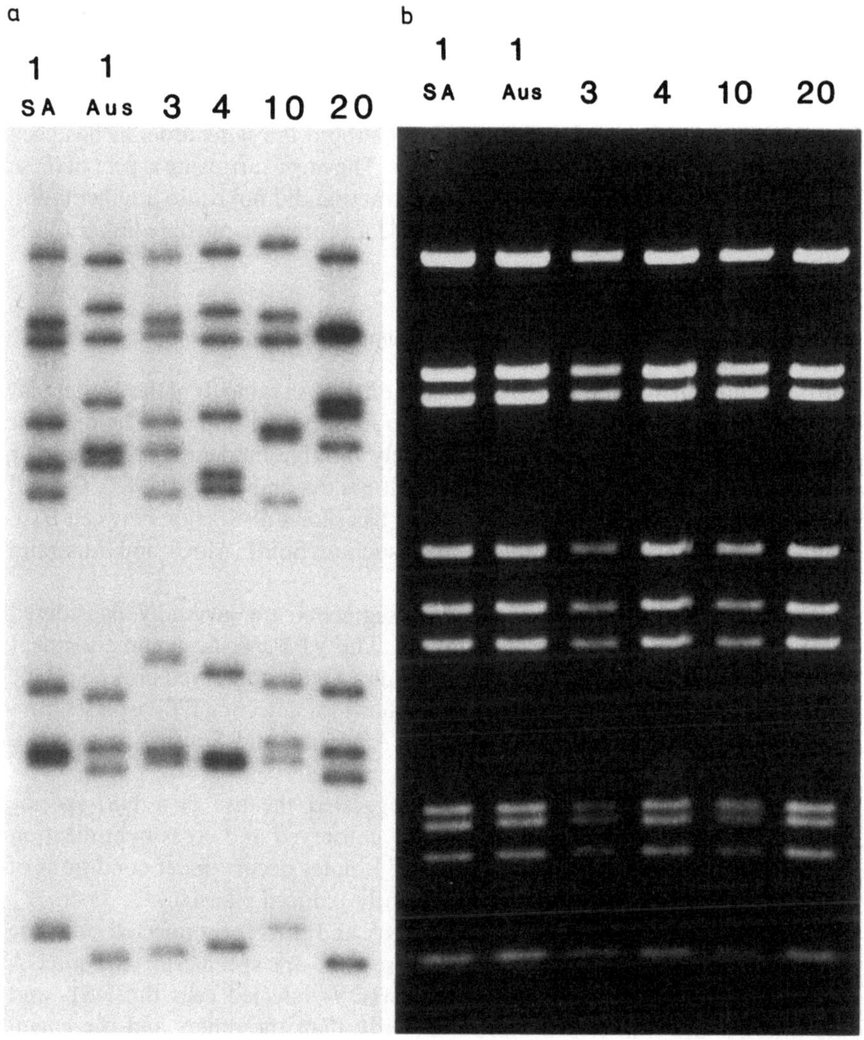

Fig. 2a, b. dsRNA profiles of BTV serotypes 1 South Africa (*1 SA*), 1 Australia (*1 Aus*), *3*, *4*, *10* and *20*. Genomic dsRNAs were extracted and separated in **a** 10% polyacrylamide (3′-end-labelled RNA, detected by autoradiography), and **b** 1% agarose gels (unlabelled dsRNA, stained with ethidium bromide). (From PEDLEY et al. (1988), with permission)

It has therefore been proposed that all references to the relative order of migration of genome segments should be standardized in terms of the migration on agarose gels (PEDLEY et al. 1988).

A question that has not been fully explored is whether the dsRNA segments are linked inside the virus particle. ELS (1973) demonstrated by electron microscopy that BTV particles liberated between six to ten fragments of dsRNA. FOSTER et al. (1978) presented evidence that the genome, when extracted under acidic conditions (pH 4.0) with SDS and phenol, is occasionally seen as an unfragmented continuous structure. In most cases, however, the genomes have a rosette configuration with loops that emanate from a central area that resembles a doughnut. These rosette patterns each contained ten loops of varying length. The relative size of each of the ten loops is of about the same order as has been reported for the individual genome segments. The most surprising aspect of these results is, however, that the SDS-phenol extraction did not cause fragmentation of the dsRNA, which would seem to exclude a protein as possible linker of the genome segments.

6.1.2 Heterogeneity of the Genome and Genomic Probes

The phenomenon of heterogeneity among the dsRNA segments of the 24 different BTV serotypes has already been referred to in Sects. 5.1 and 5.2. The different genome segments are classified as either highly variable, moderately conserved or highly conserved (KOWALIK and LI 1987; HUISMANS and CLOETE 1987). Overall nucleic acid similarity has also provided the basis for a distinction between BTV isolates from different geographical areas such as South Africa and Australia (GOULD 1987, 1988c).

The most highly conserved genome segments are generally considered suitable serogroup-specific genomic probes. The VP3-specific genome segment has been recommended by ROY et al. (1985) as a highly conserved group-specific genomic probe. The probe is particularly suitable for detecting BTV isolates from the same geographical group such as the American and South African serotypes.

HUISMANS and CLOETE (1987) have suggested the use of a NS1-specific genomic probe. The NS1 genome is highly conserved and cross-hybridization between South African and Australian BTV isolates occurs under conditions of high stringency, even though at a significantly reduced efficiency.

For the detection of virus-specific dsRNA and mRNA in infected cells, the most suitable probes are likely to be those that are specific for the mRNA species that predominate. In BTV- and EHDV- infected cells the NS1- and NS2-mRNAs are transcribed more frequently than the others and the corresponding genomic probes were consequently found to be the most sensitive in in situ hybridization (VENTER and HUISMANS, unpublished results). The same observation was made in the case of EHDV and equine encephalosis virus (VILJOEN, NEL and HUISMANS, unpublished results).

The best serotype-specific probe is segment 2 which encodes VP2. The level of interserotype cross-hybridization observed with this probe has also been shown to correlate with the low level of cross-neutralization between different serotypes (HUISMANS and CLOETE 1987). However, as pointed out by GOULD (1988a) this may also only apply to isolates from the same geographical group. For example, the genomic probes specific for the genes that encode outer capsid proteins VP2 and VP5 of BTV-1 (Aus) did not hybridize to cognate genes of BTV-1 (SA).

Similar results were obtained with the BTV-9 and BTV-20 serotypes from Australia.

6.2 Messenger RNA

The dsRNA genome segments serve as templates for the synthesis of ten single-stranded mRNA species. The mRNAs have a heterogeneous size distribution on sucrose gradients with estimated S values ranging from 12 to 22 (VERWOERD and HUISMANS 1972). BTV mRNA has also been fractionated on 3% low melting agarose gels (VAN DIJK and HUISMANS 1988) which resulted in resolution of eight of the ten BTV mRNA species. These results confirmed an earlier observation that the ten mRNA species are not synthesized at the same rate (HUISMANS and VERWOERD 1973). The synthesis of mRNA is effected by a core particle-associated RNA polymerase (transcriptase).

6.2.1 The Viral Transcriptase and mRNA Synthesis

The transcriptase is activated by removal of the outer capsid layer of the virus. This can be achieved in the case of BTV by one of four methods, namely centrifugation of virions on CsCl gradients at pH 7.0 (VERWOERD et al. 1972), treatment of virions with chymotrypsin and magnesium (VAN DIJK and HUISMANS 1980), isolation of naturally occurring core particles in BTV-infected cells (MARTIN and ZWEERINK 1972; HUISMANS et al. 1987d) and treatment of virions with 1.0 *M* $MgCl_2$ at pH 6.5 (HUISMANS et al. 1987b).

The mechanism by which the removal of the outer capsid layer activates the transcriptase is not clear. The most likely explanation is that this modification allows free access of the nucleoside triphosphates to the genome and the unimpaired extrusion of the newly synthesized mRNA. This hypothesis is supported by the observation that the BTV transcriptase is very rapidly inhibited during the course of an in vitro transcriptase assay at high core concentrations. At high concentrations the core particles tend to aggregate very strongly (MERTENS et al. 1987a; VAN DIJK and HUISMANS 1987). In the precipitated complexes the physical access to and from the core particles could be impaired with the result that mRNA synthesis is blocked.

VERWOERD and HUISMANS (1972) have reported that in vitro the BTV transcriptase has a low temperature optimum at 28°C. This is much lower than the 47°–52°C optimum reported for reovirus (KAPULER, 1970). However, the preference for a low temperature is not an intrinsic characteristic of the BTV transcriptase itself but can be explained by the observation that a reduction in temperature counteracts the inhibitory effect of high core concentration on the transcriptase reaction (VAN DIJK and HUISMANS 1980; VAN DIJK and HUISMANS 1982; VAN DIJK and HUISMANS 1987). At very low core concentrations the reaction is as efficient at 37°C as at the lower temperatures. The inhibitory effect can also be reversed by the addition of viscous compounds such as sucrose and glycerol (VAN DIJK and HUISMANS 1987). One possible explanation is that the reduction in temperature and the increase in viscosity prevent the aggregation of core particles which could inhibit the transcriptase reaction.

The RNA polymerase itself has not been characterized. ROY et al. (1988) have tentatively assigned the polymerase to VP1 on account of the degree of amino acid similarity of the latter to a variety of prokaryotic and eukaryotic RNA polymerases. More than one protein may, however, be involved in transcription. Indications have also been found that the structural integrity of the core particle is essential for maintaining transcriptase activity (HUISMANS et al. 1987d).

Analysis of the relative molar proportion in which the ten mRNA species of BTV are transcribed was carried out by analysis of mRNA/dsRNA hybrids on polyacrylamide gels (HUISMANS and VERWOERD 1973) and by direct fractionation of the in vitro synthesized mRNA by agarose gel electrophoresis (VAN DIJK and HUISMANS 1988). It was found that the different mRNA species are not transcribed at a frequency that is inversely proportional to their molecular weight as has been reported for reovirus (SKEHEL and JOKLIK 1969) and cytoplasmic polyhedrosis virus (SMITH and FURUICHI 1980). Segment 5, which codes for nonstructural protein NS1, is transcribed at more than double the predicted rate whereas segment 10, which codes for NS3, is transcribed at about one-half the predicted frequency (HUISMANS and VERWOERD 1973). It is interesting that NS1 is also translated in much higher relative amounts and comprises about 25% of the virus-specified proteins in infected cells. Nonstructural protein NS3, on the other hand, is synthesized in such small amounts that it is hardly detectable in infected cells (VAN DIJK and HUISMANS 1988). The lower than predicted level of segment 10 transcription is, however, not sufficient to explain the very low level of NS3 synthesis.

The molar ratio in which the different mRNAs are transcribed remains the same throughout the infection cycle and the ratio of mRNAs synthesized in vivo and in vitro is also identical (HUISMANS and VERWOERD 1973). The observed transcription control thus appears to be an intrinsic property of the core particle. A similar observation was made for other orbiviruses such as EHDV (HUISMANS et al. 1979).

6.2.2 mRNA Capping

It is presumed that the mRNAs of BTV are capped during transcription as in the case of reovirus and cytoplasmic polyhedrosis virus (SHATKIN 1976) and rotavirus (IMAI et al. 1983). Guanylyl transferase, a key enzyme in cap synthesis, has been assigned to core polypeptide lambda 2 of reovirus (SHATKIN et al. 1983). In the case of BTV, the function has been provisionally assigned to the core polypeptide VP4 (P.P.C. MERTENS, personal communication).

7 Conclusion

Significant progress has been made during the past 20 years in the characterization of the BTV structural components. The major aim of these investigations remains an explanation of the different biological characteristics of the virus in terms of its nucleic acids and proteins.

An important contribution to the study of the structural components was made by the cloning and sequencing of the ten dsRNA genome segments of BTV (Chap. 3). These results have provided new insight into the genetic variation of the genome segments in the BTV serogroup and have provided detailed information about the amino acid sequence of the ten virus-specified proteins.

Much of the research on the viral proteins has been focussed on immunologically important proteins such as VP2 which is involved in the induction of a protective immune response. Some progress has been made but much remains to be elucidated, particularly so with regard to the identification of the epitopes on the different viral proteins that contribute to the cell-mediated and humoral protective immune response. In this respect the structural relationship between the different proteins and the conformation of the proteins in the capsid layers of the virus is likely to be of fundamental importance.

Although some progress has been made in assigning specific functions to a few of the core polypeptides, the role of the majority of these proteins in viral replication remains largely unknown. The same applies to the nonstructural proteins. Important aspects that remain to be investigated include the association of nonstructural proteins with viral inclusion bodies, virus particles and viral components such as mRNA.

The ability to express the cloned genome segments of BTV in a variety of host cells should provide a means of investigating them in greater detail. It should also provide the opportunity of evaluating different approaches in the development of recombinant vaccines.

References

Almeida JD, Bradburne AF, Wreghitt TG (1979) The effect of sodium thiocyanate on virus structure. J Med Virol 4: 269–277

Appleton JA, Letchworth GJ (1983) Monoclonal antibody analysis of serotype-restricted and unrestricted bluetongue viral antigenic determinants. Virology 124: 286–299

Bellamy AR, Shapiro RE, August JT, Joklik WK (1967) Studies on reovirus RNA. I. Characterization of reovirus genome RNA. J Mol Biol 29: 1–17

Borden EC, Shope RE, Murphy FA (1971) Physicochemical and morphological relationships of some arthropod-borne viruses to bluetongue virus—a new taxonomic group. Physicochemical and serological studies. J Gen Virol 13: 261–271

Boyle JF, Holmes KV (1986) RNA-binding proteins of bovine rotavirus. J Virol 58: 561–568

Bremer CW (1976) A gel electrophoretic study of the protein and nucleic acid components of African horsesickness virus. Onderstepoort J Vet Res 43: 193–200

Cowley JA, Gorman BM (1987) Genetic reassortants for identification of the genome segment coding for the bluetongue virus hemagglutinin. J Virol 61(7): 2304–2306
Cowley JA, Gorman BM (1989) Cross-neutralization of genetic reassortants of bluetongue virus serotypes 20 and 21. Vet Microbiol 19: 37–51
De Villiers E-M (1974) Comparison of the capsid polypeptides of various bluetongue virus serotypes. Intervirology 3: 47–53
Eaton BT, Hyatt AD (1989) Association of bluetongue virus with the cytoskeleton. Subcell Biochem 15: 229–269
Eaton BT, Hyatt AD, White JR (1987) Association of bluetongue virus with the cytoskeleton. Virology 157: 107–116
Eaton BT, Hyatt AD, White JR (1988) Localization of the nonstructural protein NS1 in bluetongue virus-infected cells and its presence in virus particles. Virology 163: 527–537
Els HJ (1973) Electron microscopy of bluetongue virus RNA. Onderstepoort J Vet Res 40: 73–76
Els HJ, Verwoerd DW (1969) Morphology of bluetongue virus. Virology 38: 213–219
Foster NM, Alders MA, Walton TE (1978) Continuity of the dsRNA genome of bluetongue virus. Curr Microbiol 1: 171–174
Fukusho A, Ritter GD, Roy P (1987) Variation in the bluetongue virus neutralization protein VP2. J Gen Virol 68: 2967–2973
Ghiasi H, Purdy MA, Roy P (1985) The complete sequence of bluetongue virus serotype 10 segment 3 and its predicted VP3 polypeptide compared with those of BTV serotype 17. Virus Res 3: 181–190
Ghiasi H, Fukusho A, Eshita Y, Roy P (1987) Identification and characterization of conserved and variable regions in the neutralization VP2 gene of bluetongue virus. Virology 160: 100–109
Gorman BM, Taylor J (1985) Orbiviruses. In: Fields BN, Knipe DM, Chanock RM, Melnick, JL, Roizman B, Shope RE (eds) Virology, Raven Press New York, pp 907–925
Gorman BM, Taylor J, Walker PJ, Davidson WL, Brown F (1981) Comparisons of bluetongue type 20 with certain viruses of the bluetongue and eubenangee serological groups of orbiviruses. J Gen Virol 57: 251–261
Gould AR (1987) The complete nucleotide sequence of bluetongue virus serotype 1 RNA 3 and a comparison with other geographic serotypes from Australia, South Africa and the United States of America, and with other orbivirus isolates. Virus Res 7: 169–183
Gould AR (1988a) Conserved and non-conserved regions of the outer coat protein, VP2, of the Australian bluetongue serotype 1 virus, revealed by sequences comparison to the VP2 of North American BTV serotype 10. Virus Res 9: 145–158
Gould AR (1988b) Nucleotide sequence of the Australian bluetongue virus serotype 1 RNA segment 10. J Gen Virol 69: 945–949
Gould AR (1988c) The use of recombinant DNA probes to group and type orbiviruses. A comparison of Australian and South African isolates. Arch Virol 99: 205–220
Gould AR, Pritchard LI (1988) The complete nucleotide sequence of the outer coat protein, VP5, of the Australian bluetongue virus (BTV) serotype 1 reveals conserved and non-conserved sequences. Virus Res 9: 285–292
Gould AR, Hyatt AD, Eaton BT (1988) Morphogenesis of a bluetongue virus variant with an amino acid alteration at a neutralization site in the outer coat protein, VP2. Virology 165: 23–32
Grubman MJ, Appleton JA, Letchworth III GJ (1983) Identification of bluetongue virus type 17 genome segments coding for polypeptides associated with virus neutralization and intergroup reactivity. Virology 131: 355–366
Gumm ID, Newman JFE (1982) The preparation of purified bluetongue virus group antigen for use as a diagnostic reagent. Arch Virol 72: 83–93
Hall SJ, Van Dijk AA, Huismans H (1989) Complete nucleotide sequence of gene segment 8 encoding non-structural protein NS2 of South African bluetongue virus serotype 10. Nucleic Acids Res 17: 457
Hübschle OJB (1980) Bluetongue virus hemagglutination and its inhibition by specific sera. Arch Virol 64: 133–140
Hübschle OJB, Yang C (1983) Purification of the group-specific antigen of bluetongue virus by chromatofocussing. J Virol Methods 6: 171–178
Huismans H (1979) Protein synthesis in bluetongue virus-infected cells. Virology 92: 385–396
Huismans H, Howell PG (1973) Molecular hybridization studies on the relationships between different serotypes of bluetongue virus and on the difference between the virulent and attenuated strains of the same serotype. Onderstepoort J Vet Res 40: 93–104
Huismans H, Verwoerd DW (1973) Control of transcription during the expression of the bluetongue virus genome. Virology 52: 81–88

Huismans H, Joklik WK (1976) Reovirus-coded polypeptides in infected cells: Isolation of two native monomeric polypeptides with affinity for single-stranded and double-stranded RNA respectively. Virology 70: 411–424
Huismans H, Els HJ (1979) Characterization of the tubules associated with the replication of three different orbiviruses. Virology 92: 397–406
Huismans H, Bremer CW (1981) A comparison of an Australian bluetongue virus isolate (CSIRO 19) with other bluetongue virus serotypes by cross-hybridization and cross-immune precipitation. Onderstepoort J Vet Res 48: 59–67
Huismans H, Erasmus BJ (1981) Identification of the serotype-specific and group-specific antigens of bluetongue virus. Onderstepoort J Vet Res 48: 51–58
Huismans H, Cloete M (1987) A comparison of different cloned bluetongue virus genome segments as probes for the detection of virus-specified RNA. Virology 158: 373–380
Huismans H, Bremer CW, Barber TL (1979) The nucleic acid and proteins of epizootic haemorrhagic disease virus. Onderstepoort J Vet Res 46: 95–104
Huismans H, Van der Walt NT, Cloete M, Erasmus BJ (1983) The biochemical and immunological characterization of bluetongue virus outer capsid polypeptides. In: Compans RW, Bishop DHL (eds) Double-stranded RNA viruses. Elsevier New York, pp 165–172
Huismans H, Cloete M, Le Roux A (1987a) The genetic relatedness of a number of individual cognate genes of viruses in the bluetongue and closely related serogroups. Virology 161: 421–428
Huismans H, Van der Walt NT, Cloete M, Erasmus BJ (1987b) Isolation of a capsid protein of bluetongue virus that induces a protective immune response in sheep. Virology 157: 172–179
Huismans H, Van Dijk AA, Bauskin AR (1987c) In vitro phosphorylation and purification of a nonstructural protein of bluetongue virus with affinity for single-stranded RNA. J Virol 61: 3589–3595
Huismans H, Van Dijk AA, Els HJ (1987d) Uncoating of parental bluetongue virus to core and subcore particles in infected L cells. Virology 157: 180–188
Hyatt AD, Eaton BT (1988) Ultrastructural distribution of the major capsid proteins within bluetongue virus and infected cells. J Gen Virol 69: 805–815
Imai M, Akatani K, Ikegami N, Furuichi Y (1983) Capped and conserved terminal structures in human rotavirus genome double-stranded RNA segments. J Virol 47: 125–136
Inamaru S, Roy P (1987) Production and characterization of the neutralization antigen VP2 of bluetongue virus serotype 10 using a baculovirus expression vector. Virology 157: 472–479
Kahlon J, Sugiyama K, Roy P (1983) Molecular basis of bluetongue virus neutralization. J Virol 48: 627–632
Kapuler AM (1970) An extraordinary temperature dependance of the reovirus transcriptase. Biochemistry 9: 4453–4457
Kiuchi A, Rao CD, Roy P (1983) Analyses of bluetongue viral RNA sequences. In: Compans RW, Bishop DHL (eds) Double-stranded RNA viruses. Elsevier, New York, pp 55–64
Kowalik TF, Li JKK (1987) The genetic relatedness of United States prototype bluetongue viruses by RNA/RNA hybridization. Virology 158: 276–284
Lecatsas G (1968) Electron microscopic study of the formation of bluetongue virus. Onderstepoort J Vet Res 35: 139–149
Lee JW, Roy P (1986) Nucleotide sequence of a cDNA clone of RNA segment 10 of bluetongue virus (serotype 10). J Gen Virol 67: 2833–2837
Lee J, Roy P (1987) Complete sequence of the NS1 gene (M6 RNA) of US bluetongue virus serotype 10. Nucleic Acids Res 15: 7207
Letchworth GJ, Appleton JA (1983) Heterogeneity of neutralization-related epitopes within a bluetongue virus serotype. Virology 124: 300–307
Martin SA, Zweerink HJ (1972) Isolation and characterization of two types of bluetongue virus particles. Virology 50: 495–506
Mecham JO, Dean VC, Jochim MM (1986) Correlation of serotype specificity and protein structure of the five U.S. serotypes of bluetongue virus. J Gen Virol 67: 2617–2624
Mertens PPC, Sangar DV (1985) Analysis of the terminal sequences of the genome segments of four orbiviruses. Virology 140: 55–67
Mertens PPC, Brown F, Sangar DV (1984) Assignment of the genome segments of bluetongue virus type 1 to the proteins which they encode. Virology 135: 207–217
Mertens PPC, Burroughs JN, Anderson J (1987a) Purification and properties of virus particles, infectious subviral particles, and cores of bluetongue virus serotypes 1 and 4. Virology 156: 375–386

Mertens PPC, Pedley S, Cowley J, Burroughs JN (1987b) A comparison of six different bluetongue virus isolates by cross-hybridization of the dsRNA genome segments. Virology 161: 438–447
Mertens PPC, Pedley S, Cowley J, Burroughs JN, Corteyn AH, Jeggo MH, Jennings DM, Gorman BM (1989) Analysis of the roles of bluetongue virus outer capsid protein VP2 and VP5, in determination of virus serotype. Virology 170: 561–565
Oberst RD, Stott JL, Blanchard-Channell M, Osburn BI (1987) Genetic reassortment of bluetongue virus serotype 11 strains in the bovine. Vet Microbiol 15: 11–18
Oellerman RA, Els HJ, Erasmus BJ (1970) Characterization of African horsesickness virus. Arch Ges Virusforsch 29: 163–174
Palmer EL, Martin ML, Murphy FA (1977) Morphology and stability of infantile gastroenteritis virus: comparison with reovirus and bluetongue virus. J Gen Virol 35: 403–414
Pedley S, Mohamed MEH, Mertens PPC (1988) Analysis of genome segments from six different isolates of bluetongue virus using RNA-RNA hybridisation: a generalised coding assignment for bluetongue viruses. Virus Res 10: 381–390
Purdy MA, Ritter GD, Roy P (1986) Nucleotide sequence of cDNA clones encoding the outer capsid protein, VP5, of bluetongue virus serotype 10. J Gen Virol 67: 957–962
Rao CD, Kiuchi A, Roy P (1983) Homologous terminal sequences of the genome double-stranded RNAs of bluetongue virus. J Virol 46: 378–383
Ritter GD, Roy P (1988) Genetic relationships of bluetongue virus serotypes isolated from different parts of the world. Virus Res 11: 33–47
Romantschuk M, Olkkonen VM, Bamfort DH (1988) The nucleocapsid of bacteriophage Φ6 penetrates the host cytoplasmic membrane. EMBO J 7: 1821–1829
Roy P (1989) Bluetongue virus genetics and genome structure. Virus Res 13: 179–206
Roy P, Ritter Jr, GD, Akashi H, Collison E, Inaba Y (1985) A genetic probe for identifying bluetongue virus infections in vivo and in vitro. J Gen Virol 66: 1613–1619
Roy P, Fukusho A, Ritter GD, Lyon D (1988) Evidence for genetic relationship between RNA and DNA viruses from the sequence homology of a putative polymerase gene of bluetongue virus with that of vaccinia virus: conservation of RNA polymerase genes from diverse species. Nucleic Acids Res 16: 11759–11767
Shatkin AJ (1976) Capping of eukaryotic mRNAs, Cell 9: 645–653
Shatkin AJ, Sipe JD (1968) RNA polymerase activity in purified reovirus. Proc Natl Acad Sci USA 61: 1462–1469
Shatkin AJ, Furuichi Y, LaFiandra AJ, Yamakawa M (1983) Initiation of mRNA synthesis and 5′-terminal modification of reovirus transcripts. In: Compans RW, Bishop DHL (eds) Double-stranded RNA viruses. Elsevier, New York, pp 43–54
Skehel JJ, Joklik WK (1969) Studies on the in vitro transcription of reovirus RNA catalyzed by reovirus cores. Virology 39: 822–831
Smith RE, Furuichi Y (1980) Gene mapping of cytoplasmic polyhedrosis virus of silkworm by the full-length mRNA prepared under optimized conditions of transcription in vitro. Virology 103: 279–290
Spence RP, Moore NF, Nuttall PA (1984) The biochemistry of orbiviruses. Arch Virol 82: 1–18
Squire KRE, Chuang RY, Obburn BI, Knudson DL, Doi RH (1983) Rapid methods for comparing the double-stranded RNA genome profiles of bluetongue virus. Vet Microbiol 8: 543–553
Squire KRE, Chuang RY, Dunn SJ, Dangler CA, Falbo MT, Chuang LF, Osburn BI (1986) Multiple bluetongue virus-cloned genetic probes: application to diagnostics and bluetongue virus genetic relationships. Am J Vet Res 47: 1785–1788
Stanley NF (1967) Reoviruses. Br Med Bull 23: 150–154
Tokuhisa S, Inaba Y, Miura Y, Sato K (1981) Salt-dependent hemagglutination with bluetongue virus. Arch Virol 70: 75–78
Tsai KS, Karstad L (1970) Epizootic hemorrhagic disease virus of deer: an electron microscopic study. Can J Microbiol 16: 427–432
Unger RE, Chuang RY, Chuang LF, Osburn BI, Doi RH (1988a) The cloning of full-length genome segments 2, 5, 6 and 8 of bluetongue virus (BTV) serotype 17 and studies of their genetic relatedness to United States BTV serotypes. Virology 167: 296–298
Unger RE, Chuang RY, Chuang LF, Doi RH, Osburn BI (1988b) Comparison of dot-blot and Northern blot hybridizations in the determination of genetic relatedness of United States bluetongue virus serotypes. J Virol Methods 22: 273–282
Urakawa T, Roy P (1988) Bluetongue virus tubules made in insect cells by recombinant baculoviruses: expression of the NS1 gene of bluetongue virus serotype 10. J Virol 62: 3919–3927

Van der Walt NT (1980) A haemagglutination and haemagglutination inhibition test for bluetongue virus. Onderstepoort J Vet Res 47: 113–117
Van Dijk AA, Huismans H (1980) The in vitro activation and further characterization of the bluetongue virus-associated transcriptase. Virology 104: 347–356
Van Dijk AA, Huismans H (1982) The effect of temperature on the in vitro transcription reaction of bluetongue virus, epizootic haemorrhagic disease virus and African horsesickness virus. Onderstepoort J Vet Res 49: 227–232
Van Dijk AA, Huismans H (1987) The identification of factors capable of reversing the core-mediated inhibition of the bluetongue virus transcriptase. Onderstepoort J Vet Res 54: 629–633
Van Dijk AA, Huismans H (1988) In vitro transcription and translation of bluetongue virus mRNA. J Gen Virol 69: 573–581
Verwoerd DW (1969) Purification and characterization of bluetongue virus. Virology 38: 203–212
Verwoerd DW, Huismans H (1972) Studies on the in vitro and the in vivo transcription of the bluetongue virus genome. Onderstepoort J Vet Res 39: 185–192
Verwoerd DW, Els HJ, De Villiers E-M, Huismans H (1972) Structure of the bluetongue virus capsid. J Virol 10: 783–794
Verwoerd DW, Huismans H, Erasmus BJ (1979) Orbiviruses. In: Fraenkel-Conrat H, Wagner RR (eds) Comprehensive virology vol 14. Plenum, New York pp 285–345
Wade-Evans AM, Pan ZQ, Mertens PPC (1988) In vitro expression analysis of a cDNA clone of genome segment 5 from bluetongue virus, serotype 1 from South Africa. Virus Res 11: 227–240
Waldvogel AS, Stott JL, Squire KRE, Osburn BI (1986) Strain-dependent virulence characteristics of bluetongue virus serotype 11. J Gen Virol 67: 765–769
Whistler T, Newman JFE (1986) Peptide mapping of the group-specific antigens from the Australian bluetongue virus (BTV-20) and serotypes from Southern Africa and North America. Vet Microbiol 11: 13–24
Wilson L, Mesa I (1973) The mechanism of colchicine. J Cell Biol 58: 709–719

Structure of the Bluetongue Virus Genome and Its Encoded Proteins

P. Roy[1,2], J. J. A. Marshall[1], and T. J. French[1]

1 Introduction . . . 43

2 Genetic Attributes of Bluetongue Viruses . . . 44
2.1 Oligonucleotide Fingerprint Analysis of Serotypes Isolated in the United States . . . 46
2.2 Reassortment of Genome Segments . . . 47
2.3 Homologous Terminal Sequences . . . 48
2.4 Molecular Cloning of the Complete Genome of Bluetongue Virus . . . 48
2.5 Genetic Relationships Among Serotypes . . . 48
2.5.1 Genes Encoding the Outer Capsid Proteins, VP2 and VP5 . . . 49
2.5.2 The Inner Core Proteins of BTV . . . 53
2.5.3 The Nonstructural Protein Genes of BTV . . . 53
2.6 Complete Sequence of BTV Genome . . . 53
2.6.1 Segment 1, VP1 Protein . . . 56
2.6.2 Segment 2, VP2 . . . 57
2.6.3 Segment 3, VP3 . . . 62
2.6.4 Segment 4, VP4 . . . 66
2.6.5 Segment 5, VP5 . . . 66
2.6.6 Segment 6, NS1 . . . 68
2.6.7 Segment 7, VP7 . . . 68
2.6.8 Segment 8, NS2 . . . 69
2.6.9 Segment 9, VP6 . . . 69
2.6.10 Segment 10, NS3 . . . 70
2.7 Expression of BTV Genome Segments in Insect Cells Using Recombinant Baculoviruses . . . 70
2.7.1 The Baculovirus Expression System . . . 71
2.7.2 VP2 and VP5: The Outer Capsid Proteins . . . 73
2.7.3 Core Protein Morphology . . . 75
2.7.4 Virus-Coded Nonstructural Proteins . . . 78

3 Conclusions . . . 83

References . . . 83

1 Introduction

Bluetongue virus (BTV) particles were initially suggested to be morphologically similar to reovirus particles (Owen and Muntz 1966; Studdert et al. 1966; Els and Verwoerd 1969) and were subsequently confirmed to contain a double-

[1] NERC Institute of Virology, Mansfield Road, Oxford, UK
[2] University of Alabama, School of Public Health, Department of Environmental Health Sciences, University Station, Birmingham, Alabama 35294, USA

Current Topics in Microbiology and Immunology, Vol. 162

stranded RNA (dsRNA) genome as in reovirus (ELS 1973; VERWOERD 1969; VERWOERD et al. 1970; BOWNE and RITCHIE 1970). BTV, along with other morphologically related viruses, is classified as an orbivirus within the family Reoviridae (BORDEN et al. 1971; MURPHY et al. 1971). To date 24 different BTV serotypes have been identified from different parts of the world including North and South America, Australia, Africa, and Southeast Asia (HOWELL 1960, 1970; HOWELL and VERWOERD 1971; GORMAN et al. 1983; KNUDSON and SHOPE 1985). In the early 1970s VERWOERD and coworkers carried out the first biochemical studies of the BTV particle and since then many further studies on the genome of BTV have been done. In this chapter we intend to define the current state of our understanding of the genetics and the structure of the bluetongue genes and gene products.

2 Genetic Attributes of Bluetongue Viruses

The BTV genome is comprised of ten dsRNA segments, ranging in size from 0.5 to 2.7×10^6 daltons (ELS 1973; VERWOERD et al. 1979; FUKUSHO et al. 1989). The segments can be resolved by polyacrylamide gel electrophoresis (PAGE) into distinctive patterns for each serotype (Fig. 1). The RNA segments are numbered 1 to 10 in order of migration and may also be referred to as large, medium, and small segments (i.e., *L1–L3*, *M4–M6*, and *S7–S10*). Separation of the dsRNA segments by agarose gel electrophoresis, however, produces patterns that are practically identical among serotypes (PEDLEY et al. 1988), and this suggests that

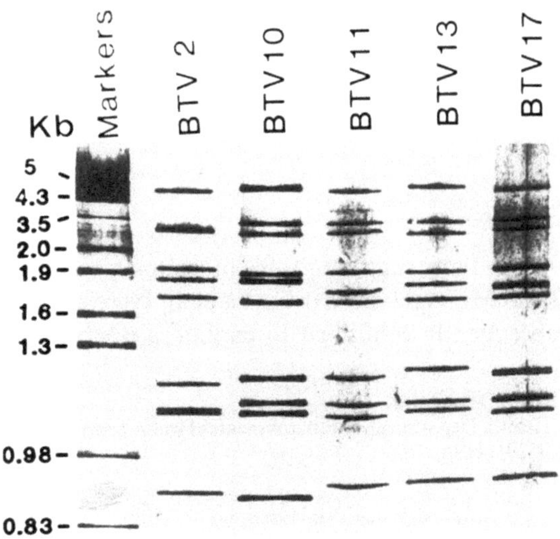

Fig. 1. Polyacrylamide gel (10%) electrophoresis of genomic dsRNA segments of five serotypes of BTV from the United States. Gel was stained with silver nitrate

Table 1. Coding arrangements of the BTV-10 genome

RNA	5′ End sequence	5′ Non-coding length	Coding region length	3′ Non-coding length	3′ End sequence
L1	GUUAAAAUGCAAUG	11	3906 VP1	37	UGA GAGCACGCGCGAGCACGCGC- CGCAUUACACUUAC
L2	GUUAAAAGAGUGUUCUACCAUG	19	2868 VP2	39	UAG GUCCUGUGACAUGGACCGGUAG- CCUCUUACACUUAC
L3	GUUAAAUUUCCGUAGCCAUG	17	2703 VP3	52	UAG AUGUGCGACCGAUCUAUGCACUUGGUA- GCGGCAGCGGAAACACACUUAC
M4	GUUAAAACAUG	8	1962 VP4	41	UAA UGCGUGACUGCUAGGUGAGGGGGGCAU- GUACAACUUAC
M5	GUUAAAAAGUGUUCUCCUACUCGCA GAAGAUG	29	1578 VP5	31	UGA ACGCAGCGGGGGAGGACCUUCCACUUAC
M6	GUUAAAAAAGUUCUCUAGUUGGCAACC ACCAAACAUG	34	1656 NS1	79	UAG UUACUGAUUUUUAGUUUUUUAUUCUUC- UUUUCAUUUCUAUUUUCUCUUAGCACUCUA- CUAGAACUUUUCAACUUAC
S7	GUUAAAAAUCUAUAGAGAUG	17	1047 VP7	92	UAG UCCACUUUGCACGGGUGUGGGUUACA.- UAUGCGGUGUGUCGGUUGUG- GGAAAUAUGUAACCCAUUUA- AACGUCUCUUAGAUUACACUUAC
S8	GUUAAAAAAUCCUUGAGUCAUG	19	1071 NS2	34	UGA CCGCAUGAUUGGGGGGGAUUUUAC- ACUUAC
S9	GUUAAAAAAUCGCAUAUG	15	984 VP6	47	UAA AGGGUCCAGGGUACCUUCUUGACGUAGG- GCGAUUUCACACUUAC
S10	GUUAAAAAGUGUCGCUGCCAUG	19	687 NS3	116	UGA GGACAGUAGGUAGAGUGGCGCCCCAAG- GUUUACGUCGUGCAGGGUGGUUGACCUCGC- GGCGUAAAUUCCCACUGCUGUAUAACGGG- GGAGGGUGCGCGAUACUACACUUAC

(Conserved 5′ and 3′ end sequences of the viral RNA segments are *underlined*. The putative translation initiation and termination codons of the mRNA sense strands are *double underlined*. The lengths of the coding and noncoding regions are shown in the center column.

equivalent genome segments from these serotypes may only have minor differences in their molecular weights, but have significant differences in their base compositions.

As summarized in Table 1, the BTV genome codes for at least seven major structural proteins (VP1–VP7) and at least three nonstructural proteins (NS1–NS3), and each of these is coded for by a separate RNA segment of the genome (GRUBMAN et al. 1983; MERTENS et al. 1984; VAN DIJK and HUISMANS 1988; PEDLEY et al. 1988).

2.1 Oligonucleotide Fingerprint Analysis of Serotypes Isolated in the United States

Oligonucleotide fingerprint analyses of the ten dsRNA species of US BTV serotypes indicated that each RNA species was unique and therefore contained

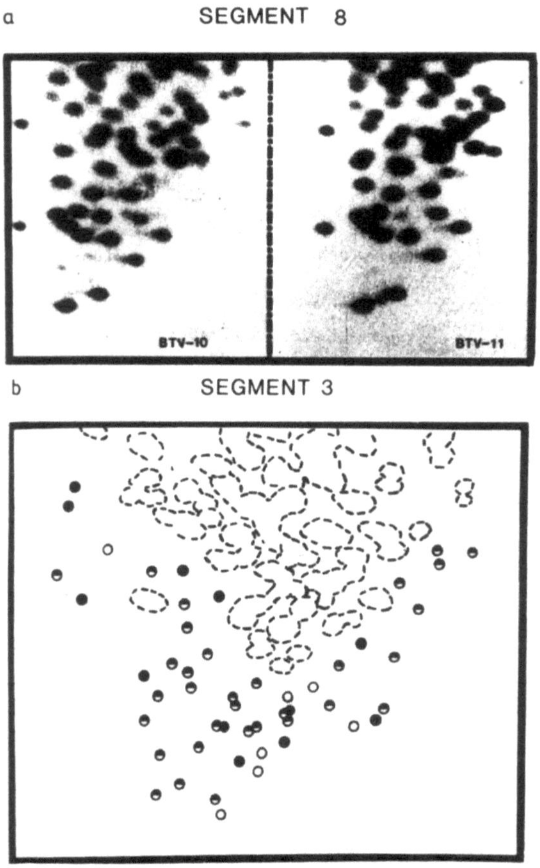

Fig. 2 a,b. RNAse T1 fingerprint analysis of different segments of BTV-10 and BTV-11. **a** Segment 8 fingerprints; **b** schematic diagram of the deduced relationships between segments 3. *Half-filled circles* represent shared oligonucleotides.

unique genetic information. However, when individual genes were compared between the serotypes (e.g., BTV-10, BTV-11, and BTV-17), the fingerprints were similar, or even identical, for most genes, but a few of the genes differed greatly. An example is presented in Fig. 2a which shows that fingerprints of segment 8 of prototype BTV-10 and BTV-11 are very similar, if not identical. Similarly, fingerprints of segment 3 of prototype BTV-10 and an alternate isolate of BTV-11 were very similar as shown in a comparative schematic diagram of fingerprints (Fig. 2b) (SUGIYAMA et al. 1981). When an isolate (80-26798, isolated from Washington State) of the BTV-10 virus was fingerprinted, it was evident that the segments were comparable with those of either prototype BTV-10 or BTV-11 except for segment 1 which was identical to segment 1 of an alternate BTV-11 isolate (COLLISSON and ROY 1983). When fingerprints of individual RNA segments of BTV-11 and BTV-17 were compared, the patterns were very similar for all equivalent segments indicating that few changes in the genome of BTV-11 could have led to the emergence of BTV-17, provided that some of the changes occurred in the gene coding for the antigenic polypeptide(s) (RAO et al. 1983a). In order to determine the extent of genotypic variation among BTV isolates, a study of the RNA sequences of BTV-11 isolates obtained in 1973 from different regions of the United States (USA), or from Colorado State, but in different years, from 1963 to 1975, was undertaken. The results indicated that all the BTV-11 isolates were related, albeit to various extents, to the US prototype BTV-11 strain. However, all showed some sequence differences, indicating that considerable evolution of the BTV-11 genome has occurred in the USA since the original 1962 isolation of the virus (SUGIYAMA et al. 1982).

2.2 Reassortment of Genome Segments

Since BTV has a segmented genome, reassortment of genome segments between different serotypes should be possible and has been demonstrated in tissue culture, in vertebrate hosts, and in *Culicoides* vectors (KAHLON et al. 1983; GORMAN et al. 1983; ROY et al. 1982; SAMAL et al. 1987). Reassortment in field isolates had been reported both in vertebrate hosts and *Culicoides* vectors (SUGIYAMA et al. 1981, 1982; Rao et al. 1983a; RAO and ROY 1983; COLLISSON et al. 1985; SAMAL et al. 1987). The generations of reassortants in the vector *Culicoides variipennis* (42%) is reportedly higher than in sheep (5%), the natural vertebrate host (SAMAL et al. 1987). Another genetic feature of BTV infection is that multiple serotypes of BTV have been recovered from individual infected sheep and cattle (OSBURN et al. 1981; OBERST et al. 1985; STOTT et al. 1982), indicating that antigenically distinct viruses can cohabit in a single ruminant.

By generating reassortant viruses involving two different serotypes, it had been possible to make protein coding assignments for certain genome segments. Serological and molecular analyses of such reassortant viruses had demonstrated that segment 2 codes for the serotype-specific neutralizing antigen VP2, segment 3 codes for the core protein VP3, segment 5 codes for the other outer capsid protein VP5, and segment 10 codes for the nonstructural protein NS3 (Roy et al. 1982; Kahlon et al. 1983). These assignments were subsequently confirmed by in vitro translation of individual RNA segments (Grubman et al. 1983; Mertens et al. 1984; Sangar and Mertens 1983; Pedley et al. 1988).

2.3 Homologous Terminal Sequences

As a preliminary step in the cloning and sequencing of BTV dsRNA segments the 3′ terminal sequences of the ten genome segments of BTV serotypes 10, 11, and 17 were determined (Rao et al. 1983b). The terminal sequence of one strand was $_{HO}$CAAUUU...5′ and the other was $_{HO}$CAUUCACA...5′ for all segments of the two serotypes. Beyond these common termini, the sequences varied considerably among segments. This was substantiated by Mertens and Sangar (1985) who reported six similar conserved terminal residues in BTV serotypes 1 and 20 and in a related orbivirus, Ibaraki virus (a member of the epizootic hemorrhagic disease, EHD, serogroup). Annealing experiments with in vitro synthesized mRNA of BTV serotype 1 and the single RNA strands separated from the dsRNA genome segments showed that the strand terminating with $_{HO}$CAUUCACA...5′ was of the same polarity as the mRNA (i.e., positive) and the strand terminating with $_{HO}$CAAUUU...5′ was of the opposite polarity (i.e., negative).

2.4 Molecular Cloning of the Complete Genome of Bluetongue Virus

To obtain full-length DNA clones representing the entire genome of BTV serotype 10, the dsRNA segments were first isolated by agarose gel electrophoresis. Complete cDNA copies of each segment were then synthesized and cloned into the *Hin*dIII or the *Pst*I restriction site of dG-tailed pBR322 plasmid (Purdy et al. 1985; Ghiasi et al. 1985; Lee and Roy 1986, 1987; Yu et al. 1987, 1988a; Roy et al. 1988; Fukusho et al. 1989). In the case of segment 5, coding for VP5 (see below), RNA-cDNA hybrid duplexes were successfully ligated directly into the *Pst*I site of pBR322 (Purdy et al. 1986). Using these techniques our laboratory has been successful in obtaining full-length clones for all segments of BTV-10 as well as a number of clones representing various RNA segments of other serotypes.

2.5 Genetic Relationships Among Serotypes

Hybridization techniques have been extensively employed to determine the genetic relationships of dsRNA genome segments between different serotypes of BTV. Huismans and Howell (1973) used mRNA-dsRNA hybridization, followed by analysis of the duplexes by gel electrophoresis, to study the relationships among South African isolates. Their results indicated that cross-hybridization occurred between the different isolates for all the dsRNA segments except for segments 2 and 5, which code for the two outer capsid proteins. A similar pattern of cross-hybridization between segments was seen in three Australian BTV isolates (GORMAN et al. 1983). However, when cross-hybridization studies involved viruses from different continents, for example, Australian serotypes with South African serotypes (4 and 10) or US serotype 17, the homology between equivalent segments was less than 30% (HUISMANS and BREMER 1981; GORMAN et al. 1981). Thus, it has been suggested that although BTV serotypes evolved from a common ancestor, the gene pools of viruses on different continents have diverged significantly as they evolved.

More recent studies have used different hybridization techniques to study the genetic relationships among five US serotypes (BTV-2, BTV-10, BTV-11, BTV-13, and BTV-17) (SQUIRE et al. 1985; KOWALIK and LI 1987). The conclusions of these studies were the same as those discussed above: RNA segments 2 and 5 showed little or no cross-hybridization between serotypes, while the other eight RNA segments, in similar studies, showed cross-hybridization, some segments more than others (e.g., segment 7 was the weakest, while segments 6 and 8 were strongest).

Northern blot analysis can reflect the extent of homology between probe and target nucleic acids under suitable hybridization conditions. We have used this technique with all available complete clones of the BTV-10 genome segments and the segment 2 clones of six different serotypes to study the genetic relationships among 20 different virus isolates from different continents (ROY et al. 1985; RITTER and ROY 1988). Hybridization conditions (e.g., temperature, ionic strength, probe concentration, and washing stringency) were chosen such that target sequences of less than 70% homology would be negative. Results of these studies are discussed below.

2.5.1 Genes Encoding the Outer Capsid Proteins, VP2 and VP5

For segment 2 of BTV-1 and BTV-2, hybridization was only observed with homologous RNAs under these conditions (Fig. 3a, RITTER and ROY 1988). Segment 2 of BTV-13 did not hybridize with any other RNA of US serotypes but did so with segment 2 of BTV-16 (an isolate from Pakistan), indicating that these two serotypes are probably more closely related to each other than BTV-13 is to the other US serotypes (Fig. 3b). Weak cross-hybridization occurred between segment 2 of the US serotypes BTV-10, BTV-11, and BTV-17 as well as the South African serotypes 4 and 15 and the Australian serotype 20 indicating some probable relationship between these serotypes (Fig. 3a, b). Similarly, hybridization between segments 2 and 5 of BTV-4 and BTV-20 has been detected by RNA-RNA hybridization (MERTENS et al. 1987). HUISMANS and CLOETE (1987), using dot-blot techniques, have also reported a low level of hybridization between segment 2 of BTV-4 and that of serotypes 10, 11, 17, and 20.

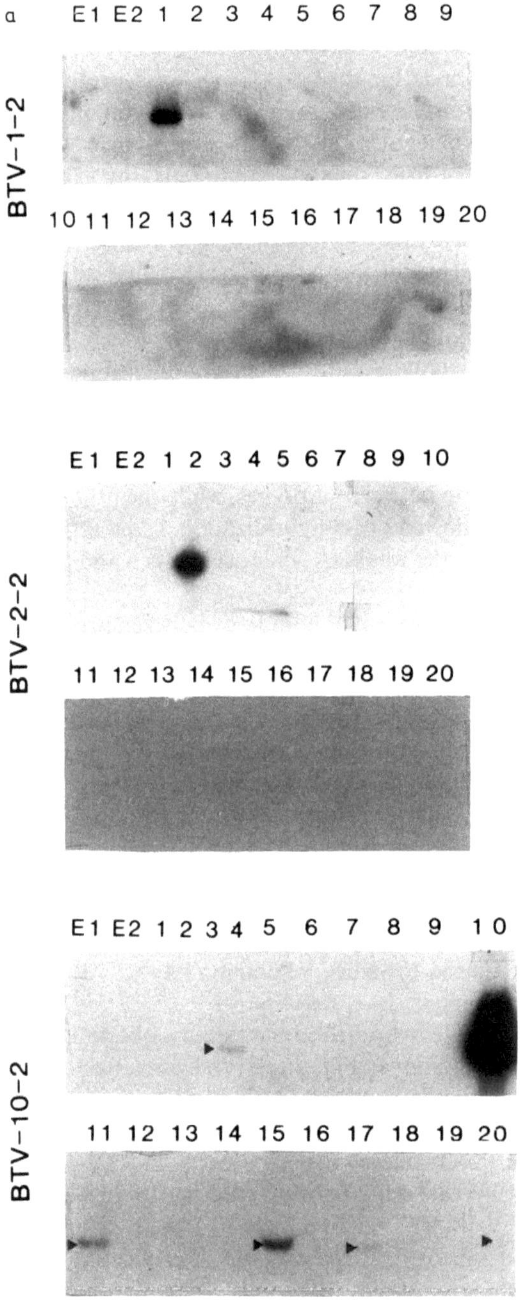

Fig. 3 (*Continued*)

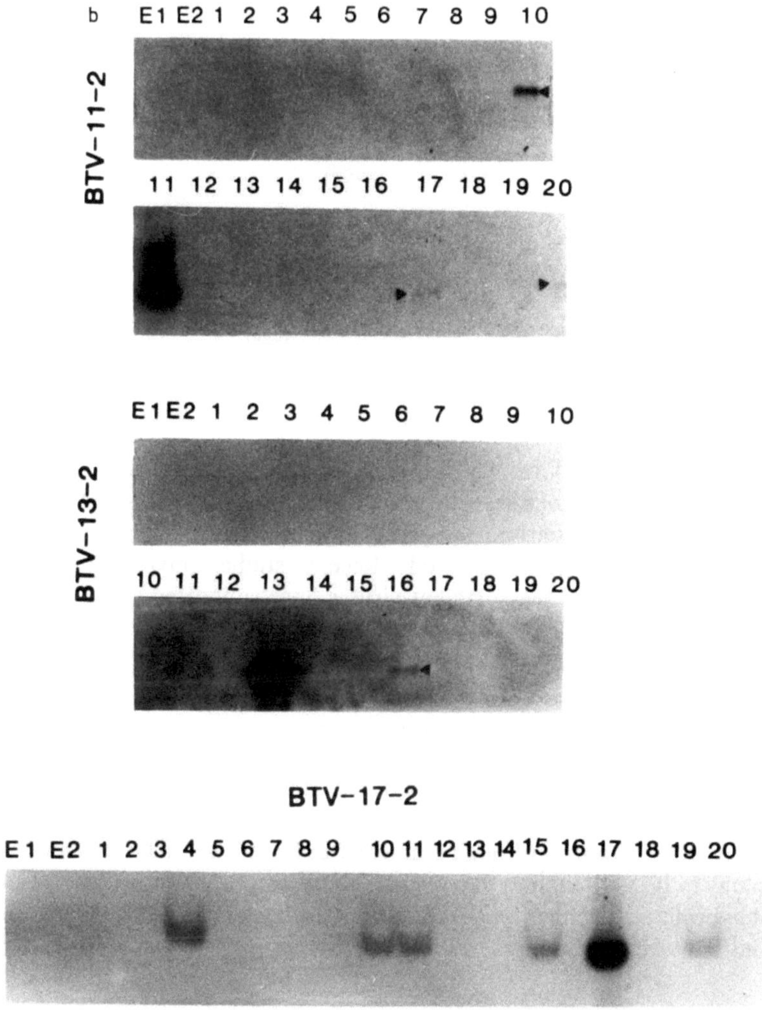

Fig. 3 a, b. Hybridization studies with the genes encoding outer capsid protein *VP2* gene of different BTV serotypes. **a** Autoradiograms of nick-translated ^{32}P-labelled DNA probes representing RNA segment 2 of BTV-1, BTV-2, and BTV-10, hybridized to genomic RNA of different serotypes. **b** Autoradiograms of similar hybridizations involving segment 2 DNA clones of BTV-11, BTV-13, and BTV-17 with different viral RNA species. The numbers (BTV-1 to -20) and E1 and E2 (EHDV 1 and 2) at the *top* of the figure represent the RNA species of different viruses. The faint bands are highlighted with *arrow heads*

Northern blots using segment 5 of BTV-10, which codes for the second outer capsid protein VP5, showed that this segment also varies among serotypes, however, for this segment US serotypes 10, 11, and 17 and S. African serotypes 3, 4, 5, and 15 are more closely related to each other than to other serotypes (Fig. 4). Thus, it is possible that VP5 plays some role in the determination of virus

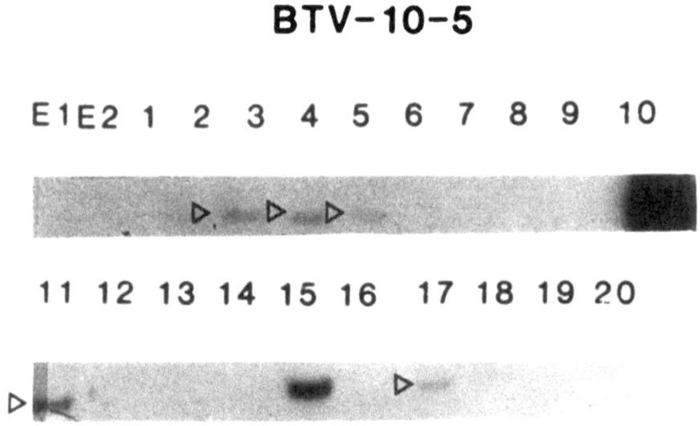

Fig. 4. Hybridization studies with the RNA encoding outer capsid protein *VP5* gene of BTV-10 (RITTER and ROY 1988)

serotype. HUISMANS and CLOETE (1987) observed some hybridization of segment 5 of BTV-4 with the homologous segments of all serotypes except 16 and 19. This discrepancy with the data discussed above may have been due to differences in the hybridization/washing conditions used. Recent studies involving cross-neutralization tests of BTV reassortment viruses have supported the idea that VP5 may play a minor role in the determination of virus serotype (MERTENS et al. 1989; COWLEY and GORMAN 1989).

2.5.2 The Inner Core Protein Genes of BTV

Hybridization experiments confirmed that segments 1, 3, and 4 are highly conserved among all BTV serotypes (Fig. 5). Segments 7 and 9 are also well conserved among serotypes, although segment 7 of BTV-7 and BTV-19 showed less homology to BTV-10 than to the other serotypes (Fig. 5). Segment 9 of BTV-10 did not hybridize with either BTV-16 (Pakistan isolate) or BTV-20 (Australian isolate) and only weakly with several other serotypes (Fig. 5; RITTER snd ROY 1988).

2.5.3 The Nonstructural Protein Genes of BTV

Similarly, hybridization studies indicated that segments 6 and 10 are highly conserved among serotypes (Fig. 6). Segment 8 is less well conserved since hybridization was weak with segment 8 of BTV-16 and BTV-20 (Fig. 6; RITTER and ROY 1988). A similar result for segment 8 of BTV-10 and BTV-16 was reported by HUISMANS and CLOETE (1987).

In summary, those segments which code for the inner core and nonstructural proteins of BTV appear to be well conserved among serotypes. However, segment 7, which codes for the group-specific protein VP7 (HUISMANS and ERASMUS 1981), and segment 9, which codes for VP6, appear to be less well conserved among serotypes than the remaining segments coding for the three inner core proteins (HUISMANS and CLOETE 1987; RITTER and ROY 1988; MERTENS et al. 1987).

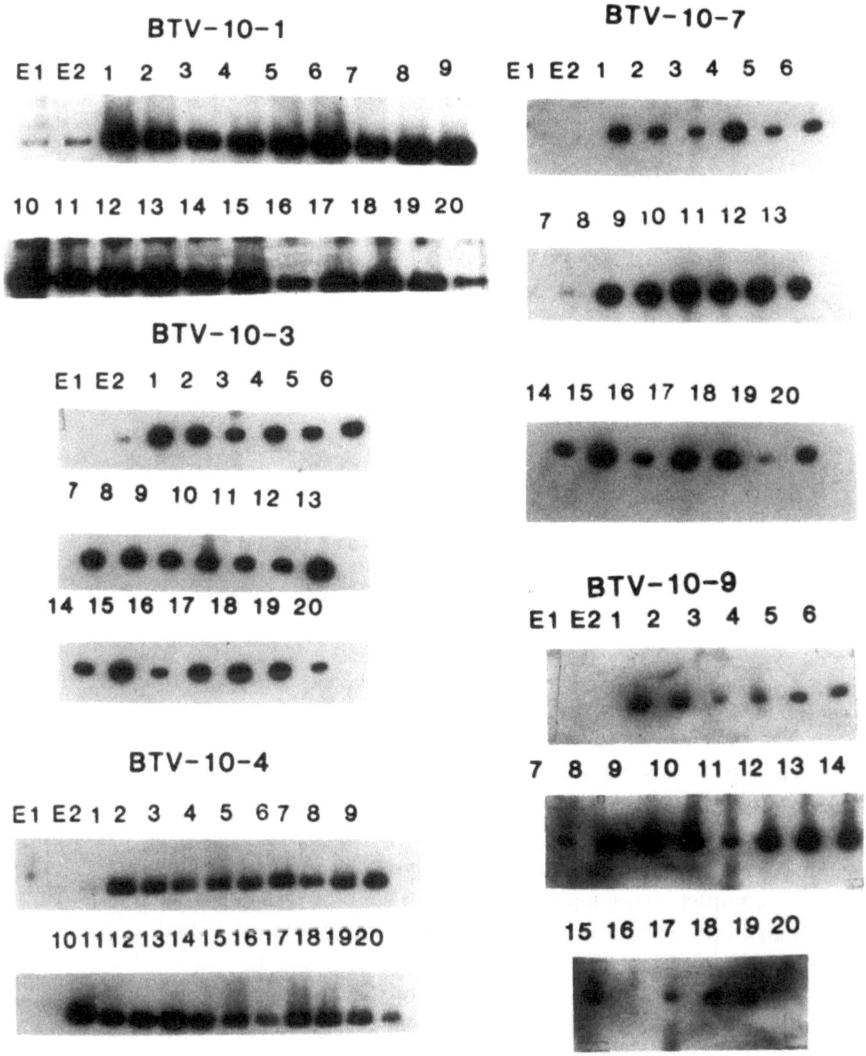

Fig. 5. Northern blot hybridization of RNA segments that encode BTV inner capsid proteins (RITTER and ROY 1988)

2.6 Complete Sequence of BTV Genome

The genome of BTV-10 has now been completely sequenced. Not only is this the first completed genome sequence of a member of the Reoviridae but also the largest reported complete genome sequence of an RNA virus. In addition, the sequences of a number of RNA segments of other serotypes have been reported. It is therefore possible to analyse the structural features of the ten segments and their predicted gene products, as well as to make comparisons between analogous

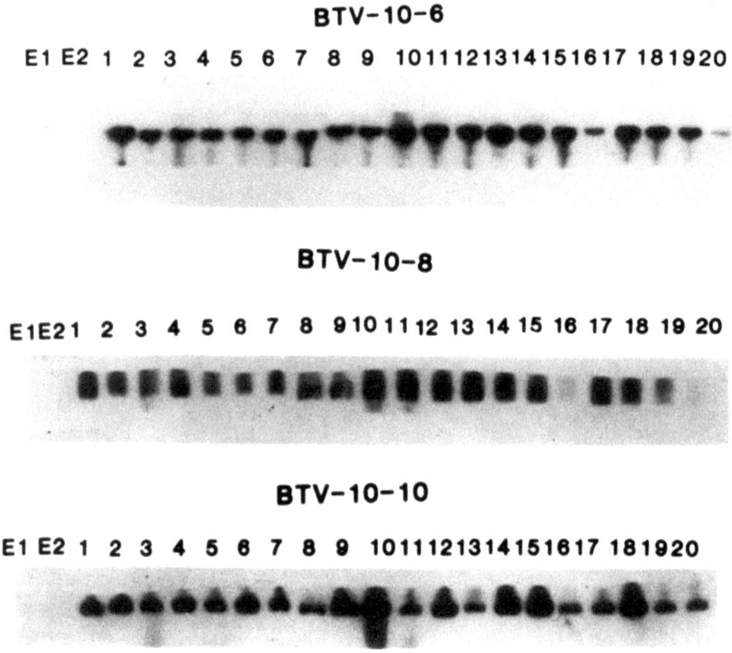

Fig. 6. Northern blot hybridization of RNA segments encoding three NS proteins of BTV (RITTER and ROY 1988)

genes and products from different serotypes (PURDY et al. 1984; FUKUSHO et al. 1987; GHIASI et al. 1985, 1987; GOULD 1987, 1988a, b; GOULD and PRITCHARD 1988; GOULD et al. 1988; HALL et al. 1989; WADE-EVANS et al. 1988; YAMAGUCHI et al. 1988a, b). The results of these analyses are discussed below.

The complete sequence of the BTV-10 genome was determined to be 19218 base pairs (bp) long (13×10^6 daltons). The sizes of the individual segments vary from 3954 bp (segment 1, 2.7×10^6 daltons) to 822 bp (segment 10, 5×10^5 daltons). The overall base composition of the genome is 28.1% A + U; 21.9% G + C; the base compositions of the individual segments are all quite similar and range from 24.6% to 20.7% G + C. The 5′ noncoding regions range from 8 bp (segment 4) to 34 bp (segment 6) in length, while the 3′ noncoding regions are longer, ranging from 31 bp (segment 5) to 116 bp (segment 10).

Presented in Table 1 are the summarized data for the coding arrangements of all ten of the dsRNA segments of BTV-10. Sequencing of the cloned segments confirmed the presence of conserved terminal sequences (see above) on all ten segments (RAO et al. 1983b). Each RNA segment has highly conserved terminal sequences (i.e., 5′-GUUAAA-----3′ on the 5′ end and 5′----CACUUAC-3′ on the 3′ end of the positive-sense strands). In addition, other features proximal to these conserved sequences can be recognized in most of the segments. For example, nine of the ten segments have another A following the 5′ conserved sequence (i.e., at position 7) and six of the segments have two A residues at positions 7 and 8. At

Table 2. Predicted amino acid compositions of the ten primary gene products of BTV-10

Amino acid	VP1 Core	*n*	VP2 Shell	*n*	VP3 Core	*n*	VP4 Core	*n*	VP5 Shell	*n*	NS1	*n*	VP7 Core	*n*	NS2	*n*	VP6 Core	*n*	NS3	*n*
Alanine (A)	76	58	53	55	3	70	40	61	49	93	41	74	40	115	21	59	31	95	20	87
Arginine (R)	91	70	65	68	69	77	54	83	28	53	43	78	24	69	27	76	32	98	11	48
Aspartic (D)	77	59	69	72	59	65	48	73	26	49	32	58	13	37	29	81	18	55	11	48
Asparagine (N)	53	41	41	43	41	46	24	37	16	30	20	36	18	52	12	34	9	27	8	35
Cysteine (C)	12	9	16	17	5	6	6	9	3	6	16	29	3	9	7	20	1	3	2	9
Glutamic (E)	84	65	60	63	50	55	48	73	55	105	39	71	14	40	34	95	36	110	15	66
Glutamine (Q)	42	32	37	39	42	47	8	12	19	36	26	47	20	57	16	45	9	27	10	44
Glycine (G)	69	53	49	51	43	48	33	50	33	63	32	58	23	66	20	56	40	122	6	26
Histidine (H)	21	16	29	30	16	18	27	41	17	32	14	25	6	17	4	11	6	18	3	13
Isoleucine (I)	95	73	70	73	65	72	38	58	48	91	39	71	24	69	19	53	19	58	11	48
Leucine (L)	111	85	90	94	82	92	63	96	43	82	42	76	28	80	21	59	16	49	23	100
Lysine (K)	87	67	61	64	27	30	36	55	40	76	23	42	1	3	29	81	30	91	19	83
Methionine (M)	50	38	20	21	35	39	27	41	17	32	23	42	19	54	13	36	6	18	13	57
Phenylalanine (F)	65	50	43	45	39	43	25	38	20	38	26	47	14	40	11	31	2	6	9	39
Proline (P)	51	39	36	38	47	52	32	49	15	29	20	36	21	60	17	48	7	21	11	48
Serine (S)	83	64	47	49	42	47	37	57	31	59	25	45	11	32	20	56	20	61	21	92
Threonine (T)	78	60	48	50	53	59	24	37	22	42	23	42	30	86	13	36	18	55	15	66
Tryptophan (W)	11	8	13	14	10	11	15	23	2	4	13	24	5	14	5	14	3	9	1	4
Tyrosine (Y)	56	43	47	49	39	43	28	43	14	27	28	51	9	26	10	28	4	12	3	13
Valine (V)	90	69	62	65	73	81	41	63	28	53	27	49	26	74	29	81	21	64	17	74
Total	1302		956		901		654		526		552		349		357		328		229	
Net charge	+27.5		+11.5		−5.0		+7.5		−4.5		+2		+1		−3		+11		+5.5	
Size	149588		111023		103326		76433		59163		64445		38548		40999		35750		25602	

The individual composition, predicted size and charge, at neutral pH, of each gene product is given. For comparative purposes the amino acid contents are also expressed as the number (*n*) per 1000 residues. The presumed location of each protein in the virion is given (outer capsid. Shell; inner capsid, Core; nonstructural NS)

the 3′ ends of the positive-sense strands the conserved sequence is preceded by AC in 7 of the 10 segments, and in the remaining 3 segments (4, 5, and 6) it is preceded by CA or CC. Also UU dimers are found proximal to the 3′ ends in seven of the ten segments (1, 2, 5, 6, 7, 8 and 9). The 3′ noncoding regions of seven of the segments (3, 4, 5, 7, 8, 9, and 10) are purine-rich; it is not known if this plays any role in transcription, translation, or morphogenesis.

Apart from segment 1 the first AUG codon on the positive RNA strand of each of the segments initiates a long open reading frame. Segment 1 has an additional AUG codon upstream of the codon which initiates the open reading frame (residues 7–9); it is not known if this affects the translational efficiency of the gene. Similar AUG codons have been reported for segment 1 of BTV-1, BTV-11, and Ibaraki virus (MERTENS and SANGAR 1985; RAO et al. 1983b). Some of the initiating AUG codons have some of the features of the consensus flanking sequences for initiating codons proposed by KOZAK (1981), but none have all the features. For example, seven of the segments (1, 2, 3, 5, 6, 7, and 8) have a G at position +4 (counting the A of AUG as +1) and all of the segments, except segment 9, have a G or an A at position −3. All three possible translation termination codons are used; four gene products terminate with UGA, four with UAG, and the remaining two with UAA.

Table 2 is a summary of the predicted amino acid composition, size, and net charge of all the primary gene products of BTV-10. The amino acid composition is also given as the number per 1000 residues; the mean composition of all the products is as follows: A, 76.7; R, 72.0; D, 59.7; N, 38.1; C, 11.7; E, 74.3; Q, 38.6; G, 59.3; H, 33.1; I, 66.6; L, 81.3; K, 59.2; M, 37.8; F, 37.7; P, 42.0; S, 56.2; T, 53.3; W, 12.5; Y, 33.5; V, 67.3. When individual gene products are compared with these values several striking variations are evident. These are discussed below, with other data, for each individual segment.

2.6.1 Segment 1, VP1 Protein

This, the largest of the segments, codes for the minor core protein VP1. The molecule is a highly basic protein with a positively charged carboxyl terminus. In fact this protein has the highest predicted net positive charge (+27.5 (R + K + 1/2H-D-E) at neutral pH), although it is no more abundant in charged residues per unit length than the other BTV gene products. VP1 is also rich in the hydrophilic residues serine and threonine and the aromatic residues phenylalanine and tyrosine compared with the other gene products (see Fig. 7). The VP1 primary sequence has some homology with that of vaccinia virus 146 kD polymerase subunit, the β-chain subunits of *Escherichia coli* and common tobacco chloroplast RNA polymerases, *Saccharomyces cerevisiae* RNA polymerase II and III, and *Drosophila* polymerase II (ROY et al. 1988). These homologies may reflect similar functional domains between DNA- and RNA-directed RNA polymerases, and there is a possibility that they may have all evolved from a common ancestral gene. Recent evidence suggests that VP1 is indeed a transcriptase component (see Sect. 2.7).

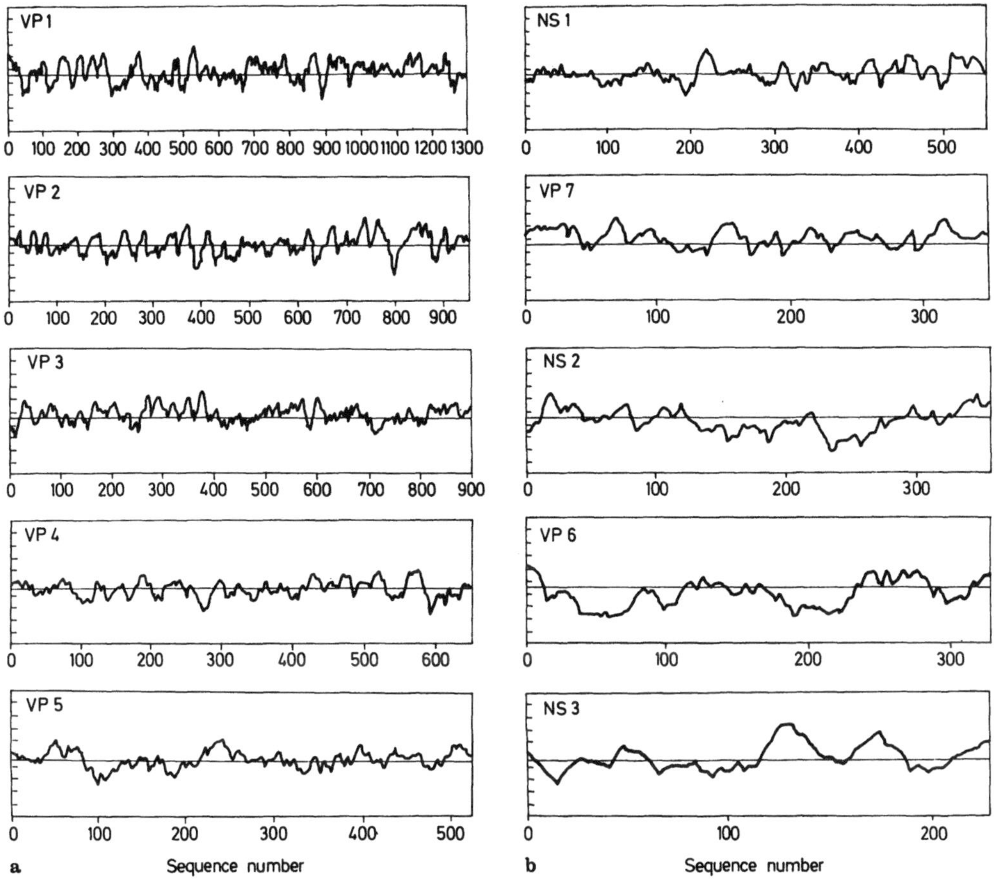

Fig. 7a,b. Hydropathic plot for the predicted ten RNA gene products of BTV-10. The regions of the predicted proteins with a net hydrophobicity (*areas above the center line*) are displayed. The plot involves a span setting of 21 amino acids

2.6.2 Segment 2, VP2

This gene product is a major component of the outer capsid of BTV and is the principal serotype-determining antigen. It elicits neutralizing antibodies (HUISMANS and ERASMUS 1981; KAHLON et al. 1983) and is the BTV hemagglutinin protein (COWLEY and GORMAN 1987). Together with the other outer capsid protein, VP5, it exhibits the greatest sequence variation of all the segments among serotypes (FUKUSHO et al. 1987; GHIASI et al. 1987; RITTER and ROY 1988; YAMAGUCHI et al. 1988a, 1988b; GOULD 1988a) The neutralizing antigen VP2 appears to be hydrophilic in nature and contains many charged residues (Fig. 7). VP2 is rich in aromatic residues and conserved cysteine residues which may indicate a highly ordered, disulfide-bonded structure.

Table 3. Terminal sequence comparisons of 5′ and 3′ sequences of six BTV *L2* genes

Serotype	Sequence			
BTV-10	5′	GUUAAAAGAGUGUUCUACCAUG.........	UAGGUCCUGUGACAUGGACCGGUAGCCUCUUACACUUAC	3′
		*************** * ****	**** ***** ****** * ********* ******	
BTV-11	5′	GUUAAAAGAGUGUUCCAUCAUG.........	UAGGUCCCUUGACACGGACCGAUGGCCUCUUAUACUUAC	3′
		************ ** ******	***** ****** * *********************	
BTV-17	5′	GUUAAAAGAGUGAUCCACCAUG..........	UAGGUCCUAUGACACGGACCGAUGGCCUCUUACACUUAC	3′
		******* * * *	*** **** *** ** * * **********	
BTV-13	5′	GUUAAAAACGCUAGCUCGAGGAUG	UAA-UCCCGUGACUGGGAUUGA-GCGCGUUUACACUUAC	3′
BTV-2	5′	GUUAAAACAGCAUCGCGAUG.............	UGACCUGCGUGACUGCAGGUCCGCGAUCUGUUCAACUUAC	3′
		******* ** ********	***** ** **** * ******** ** * ******	
BTV-1	5′	GUUAAAAUAGUGUCGCGAUG...........	UGACCCGCUUGACCGUGGGUCCGCGCUAUCAUACUUAC	3′

Three US serotypes compared with that of BTV-10 and serotype 2 compared with Australian serotype 1. The conserved sequences are indicated by *stars*. Putative translation initiation codon and the termination codons have been underlined.

In order to study the nature of antigenic variation among serotypes segment 2 of four other US serotypes (BTV-2, BTV-11, BTV-13, and BTV-17) and one Australian serotype (BTV-1) have been sequenced (FUKUSHO et al. 1987; GHIASI et al. 1987; YAMAGUCHI et al. 1988a, b; GOULD 1988a). All of these sequences contain the conserved 5′ and 3′ terminal sequences discussed above (Table 3). Segments 2 of BTV-1 and BTV-2 are 2940 bp and 2943 bp long, respectively, which are slightly longer than those of BTV-10 and BTV-11 (both 2926 bp) or BTV-13 (2935 bp) or BTV-17 (2923 bp). The 5′ noncoding region of BTV-1 and BTV-2 is 17 bp which is shorter than that of the other four serotypes (19 bp for BTV-10, BTV-11, and BTV-17; 21 bp for BTV-13). The 3′ noncoding region of BTV-10, BTV-11, and BTV-17 is 36 bp, 34 bp for BTV-13, and 35 bp for BTV-1 and BTV-2. The noncoding regions of BTV-10, BTV-11, and BTV-17 are well conserved by comparison with those of BTV-13 and BTV-2, those of BTV-2 are more homologous with those of BTV-1.

Alignment of the predicted amino acid sequences of the six serotypes shows that the VP2 proteins of BTV-10, BTV-11, and BTV-17 are closely related, while those of BTV-13 and BTV-2 are more closely related to each other and to BTV-1, the Australian isolate (Fig. 8). The close similarities of the VP2 sequences of BTV-10, BTV-11, and BTV-17 serotypes are indicated by the fact that only one gap was required for maximum homologous alignment between them, as shown in Fig. 9. Similar alignment comparisons among the VP2 of BTV-2 and BTV-13 or BTV-1 required several gaps (Figs. 10 and 11). The sequence homology does not appear to reflect the geographical distances among isolates (Table 4). In general the differences in amino acids among the six serotypes are distributed evenly throughout the whole protein (YAMAGUCHI et al. 1988b), but at least five highly variable regions (Fig. 8, underlined) can be identified. Conversely, there are a number of highly conserved regions (indicated by asterisks) which are evident; conserved amino acids together with conservative changes (on the basis of charge or polarity) account for approximately 43% homology between all six serotypes. The carboxyl terminus of VP2 appears well conserved, which may reflect some form of evolutionary constraint such as a need to interact with other BTV proteins (e.g., the other outer capsid protein VP5). Also the middle region is relatively well conserved, with the absolute conservation of an octapeptide between all six serotypes (NPYPCLRG; positions 360–367, Fig. 12), which may reflect a role in the preservation of overall protein conformation. The six cysteine residues that are conserved among all six serotypes (indicated by dots in Fig. 8) may be involved in disulfide bridges important for protein structure. The common phylogenetic origin of all six viruses was revealed by the diagonal lines when the apparently distantly related VP2 proteins (e.g., BTV-10 and BTV-1 or BTV-2 or BTV-13 and BTV-2 and BTV-1 or BTV-13, etc.; see Fig. 13) were compared although, as shown by arrows, some gaps are noticeable. Also the hydropathic profiles of the six VP2 sequences are similar (data not shown). This pattern of variable and conserved regions may be due to the need for VP2 to preserve its overall structure, which may be necessary for interactions with other

```
                    10        20        30        40        50        60        70        80        90       100
BTV-10   MEEFVIPVFSERDIPYSLLNHYPLAIQIDVKVDDEGGKHNLIKIPESDMIDVPRLSIIEALNYRPKRNDGVVVPRLLDITLRAYDNRKSAKNA-KGVEFM
BTV-11   ***********TE******S*****VRTN**IANVDEG*DVV************V**V***AAK*T****I***************D**AM*S*-R*****
BTV-17   ********Y**DE***A**SR******TN**IE*VE****VV*********I*K*T*V**M**K*A****I***************D***T*S*-R*I***
BTV-13   ***L****IT**-FDKR*VGR*DYV*ELARPEG**WSG*DVTH**DRR*F*IKVQP*RD*ID*K*VE***E*L**I**MSIAC**M**RMMKR-D**D*V
BTV-2    *D*LG**IYG-*NY*EH**KG*EFL*NTG**YPSQ**R*DVS****MFAY*IKDEG*RD**K*M*T******L**IV**S*KG**I***VIE*N*SNS*H
BTV-1    *D*LG**IYK-*GF*EH**HG*EFT*DSST*IQSV**R*DVT*L**MNAY*IKSEGMRT**W*N*V****F*L**V******G**GKRAVIDSSRHKS*H

                   110       120       130       140       150       160       170       180       190       200
BTV-10   TDTKWMKWAIDDKMDIQPLKVTLDN--HCSVNHQLFNCIVKARSANADTIYYDYYPLENGAKRCNHTNLDLLRSLTTTEMFHILQGAAYALKTTYELVAH
BTV-11   *NA*********R*********AI*D--*NA*************P*****V**S*F**RDKV*K**********G*********M******C**SS***ITN
BTV-17   *NAR********R************H--Y***********V***NA***********F***DYK****************NM*L**A******SI*SS*****Y
BTV-13   S****LE*M*Q*S**V*****DMKEDHSTVQYDMFSAKLHVDSRKADT*S*NILALETKEGAQ*H*VHTNIWNHMIRNHL**AV*ESC*IF*P**K*TVN
BTV-2    ***S*VQ*M*K*S**Q*****ISI*-EE*SR*V*S****Q**IDAKK***LS*HLEAI*DAE*A*L**RGQ*CNH**RMDLL*AA*EI***I*P**Q*IV*
BTV-1    **ER*VQ*MMK*S**A*****G**D-QTQKIA*S*H**V**ID*KK***MS*HIE*I*DSL*G*L**RTM*WNH*VRV**S*AA*EI*****P**DI*V*

                   210       220       230       240       250       260       270       280       290       300
BTV-10   SERENMSESYQVGTQRWIQLRKGTKIGYRGQPYERFISSLVQVIIKGKIPDEIRTEIAELNRIKDEWKNAAYDRTEIRALELCKILSAIGRKMLDVQEEP
BTV-11   ***N*TE*T*AP*VHNR*R*VR**R***K*EA*S**V*******R*Q*QT*P**VDD**R**E*RT*****QF*S*K********************NTH***
BTV-17   ***GSLE*T*V**QPK**H*TR**R**NS*LS*******M***SVN*******AN***Q****RA***T*T***GR*******S***T*******NTH***
BTV-13   ***RTPD*DF*I*NPQFLT**RNQQ*FLGDDA*KKTAKG****LVN*VV**I**N***A*DA*R*K**QGN*E**H*KS****NL********VNLE***
BTV-2    ***AST*DNFEL*R*DV*T**R*HRVQMGDEA*TKLMER**RLTVQ*NV*RK*QS**EQ*EA*RTR*ATGR**PAH*NSQD**R***R***I***QEA**
BTV-1    A**RDR*QPFRP*D*TL*NFSR*Q*VQMNHNS**KMVEG*AHLV*R**T*EL**D**TK*DE*CNR**RSR**PG**K*Y****V**TV******QEK**

                   310       320       330       340       350       360       370       380       390       400
BTV-10   KDEMALSTRFQFKLDEKFIRTDQEHVNIFKVGGSATDDGRFYALIAIAGTDTQQGRVWRTNPYPCLRGALIAAECELGDVYFTLRQTYKWSLRPEYGQRE
BTV-11   ****D***********D**KK**S**I***N**AP**HE*********A****R********************Q*****H***RV******QD**RT*
BTV-17   ****D*************N*A*S******G*R*P***E**********A****K*******************V**********S***RV*T*********H*
BTV-13   ***RD**L***H***D**AKN***RNV**AQKSQRN*QD***V*MV**AS**NNS***WS*****************K*****YK**SW*E**V*EG*KP*D
BTV-2    V**DS**L***RA*****RLN*S*RNK**EHKSHKK*ED***V*LR**AS**YNS*I*WS*********T*****TK*********SW*D**V*SS*IP**
BTV-1    A**AN**I***E*I*N**RQH*S*RLK**EHRNQRR*ED***I*LM**AS**FNT***WS*********T***S*TK*****SMM*LW*D**V**T*IPY*

                   410       420       430       440       450       460       470       480       490       500
BTV-10   RPLEDNKYVFARLNLFDTNLAVGDEIIHWRYEVYQPKETTHDDGYICVSQKGDDE---LLCEVDEDRYKEMFDRMIQGGWDQERFKLHNILT-EPNLLTI
BTV-11   V***N*****S*I****S**E***QVV**K**IDG*A***Y*N****KTERE*G*----*V*KIS*EK**T*L***************YSV**-D******
BTV-17   *Q**N*****N*I****S******Q*******KASA***Y*S**M*RHEAEE*---***KIN**K**E*L************************-D******
BTV-13   LDRQYE**IVG*V****LEAEP*TKVL**E**LISKLY*VSNHEGNQCDLHP*EGE--IVTKF*DT**SD*IQTI*NE**K*ND**MFKM*KD*G*P*LY
BTV-2    *ER*TE**I*SKI****YEAGPSSKV***E*QL*KRERVVTLERGNPCDLYP*EDDEVIITKF*DAK*S**VGEI*D***ND*E**MYKL*QEKG*V***
BTV-1    KSR*QE**IYG*V****YVAEP*TK****E*KLN*QIKDITYEQGNPCDLFP**DEA-IVTKF*DVA*GQ*VNDL*N*********M*K**KSQG*V***

                   510       520       530       540       550       560       570       580       590       600
BTV-10   DFEKDAYLGARSELVFPPYYDKWINSPMFNARLKIARGEIATWKADDPWSNRAVHGYIKTSAESLEYALGPYYDLRLQLFGDTLSLGQRQSAVFEHMAQQ
BTV-11   ********NI***F*L*S*F*Q**Y********R*TH***G*R*SA***NK*V*F**V*A*T**P*****Q*F*T*I**Y**A***K*S*****Q*QS**
BTV-17   ********NS*****L*D*F****S*********R*TK***G*S*K****N****R****PL****DFV*********LF*DE****K*E*****QYLS*L
BTV-13   *L***IK*DRV*RV*****F*Q*TYV*******I*PCEV*VGER*NI**YVK*THRPLKADCI*LMR*HMSQ*M***VS*Q*TS**IK*TP*SIHQSL*RD
BTV-2    *****TK*YNT**V*L*D**G***VA****SKMR*IET****N*S***MIK*TLKPMTDDPV*LQR*T*AR***I*PG*M*RS*NRT*T**TFDAKVSEL
BTV-1    ******K*TSNEGVAM*E*F****IA*****K*R*KH****QRRN***MVK*TLSPIAFAPIVLQRLT*ARF**I*PAIM*QA**RQ*G**TYD*EISKI
```

```
                 610       620       630       640       650       660       670       680       690       700
BTV-10  DDFSTLT-DYTKGRTVCPHSGGTFYTFRKVALIILSNYERLDPSLHEGREHETYMHPAVNDVFRRHVLEMKDFSQLICFVFDYIFEKHVQLRNAKEARRI
BTV-11  E**PV**-S*A**DV*******AL********MLMA*****S*D****M*DH**T**SIGGANQKRI***R*********II******R*D***DMR*****
BTV-17  ***PA**--QLR*DA*******AL********FLIG***K*S*D****M**QR*V**STGGTYQKR******SC**T***I*******RE***DT****Y*
BTV-13  ASYAEILSRRRENLDYKSQCPIVTNL*LLEKFFL*IFTTMEKHYWEMDDDETE*E**KIDPSKFEVEGTLH*V**VMVHL**RF***RRF**TVD*S*W*
BTV-2   P*YEKVVSRFGVIKKPTRPCVTLTGRYILEKYSL*LIDILKYHTEV**NPQ*EFT**RIDPQ*KFNGNTLS*LN*TVV*IV**LH**RNYV*SIY***Y*
BTV-1   EGYAEILQRRGIVQIPKKPCPTVTAQYTLERYALFLINILEQHIIQSTD*DVM*S**RVDYKLEV*GENII*I**IVI****FL**RRRTV*GVY*S*YM

                 710       720       730       740       750       760       770       780       790       800
BTV-10  IYLIQNTSGAYRLDVLREAFPNFLKHVMNLRDVKRICDLNVINFFPLLFLVQDNISYWHRQWSIPMILFDQVIRLIPVEVGAYANRFGLKSFFNFIRFHP
BTV-11  L**V*SLGEPQ*****SV*S***SRYFLK*K**Q**S*******L*****I*************AV****Y*DT*K*****************I******T****
BTV-17  V****SLT*TQ**S***ST****FQRLLM*KEI*FVR*******L**M***H****************V***DT*K*****************F***M**T****
BTV-13  LH**RSA***R**E**SRF**A*S-DGLRI*EF*KVR*IMLL**L*F***TG***A*E****AV*V*FYADK*MI**A*****Y*****TCILELMM*F*
BTV-2   *SR*RSST**A*MSIIEFY**T*ARLIS*A*EPTYVK***AL**L****I*G**MI*K********LL*YTDRVKV**L***SSN**Q*FV*YLEYMF*F*
BTV-1   VTR*RDAQ*QN*IN*IT*F**T*GY*LSRVKEATI*QEI*YL**L**F***S***I*T*K***V*LF*YAHELKV**L***S*ND*CS*V*YIEYMV*F*

                 810       820       830       840       850       860       870       880       890       900
BTV-10  GDSKKRQDADDTHKEFGSICFEYYTTTKISQGEIDVPVVTSKLDTLKLHVASLCAGLADSLVYTLPVAHPKKSIVLIIVGDDKLEPQIRSEQIVNKYYYS
BTV-11  **A****K**********LIS*N**AN***A**GVHT****T*******I*LS**********V*************C*****************HV****V*S***F*
BTV-17  *EL**K*I*E*I*****VVA*****N*******NVHT**M*T*M*V*RV*LS**********V*************C*****************HT******SR*N**
BTV-13  SYDTRNENLSEDVRACIGPIIN**LD*T**N*G*QTSI*ST*ALLYETYLS*I*G*FSEAILWY**IT**S*CLIALE*S*ALTS*EL*IDK*KRRFPL*
BTV-2   SLADRTSKV*ESMIKVSKEMVN**MK*T**E*GVNLN**ST*SLLYDIYLS*V*G*VS*GV*WY**IT**Y*CV*A*E*C**RVPARL*CDRLKLRFPL*
BTV-1   SKAFRTSKL*EVQPKIAREMLK**IN***FE*G*NLN***T*QLLYETYLA*I*G**S*GI*WY**IT**N*CL*A*E*S*ERVPAS**ASH*KLRFPL*

                 910       920       930       940       950       960
BTV-10  RRHISGVVSICVNQGGQLKVHSMGITRHRICDKSILKYKCKVVLVRMPGHVFGNDELMTKLLNV
BTV-11  *K*V*******IG*ND****Y*S**V*****E*F**R*********K***Y***************
BTV-17  *K**C*I**VTIG*NS**R**TS**VK**V***F***H****I*******Y***************
BTV-13  SN*LK*I*Q*S*RP*RTFS*VTQ**VK**V*K*TL*R*R*D*I*IQT**Y*******L*****I
BTV-2   AQ*LK*I*V*QI*EE*GFD*YTE**VT**V*K**L**HV*DI**LKFH********ML******
BTV-1   VK*LK*I*I*Q*DEE*KFT*Y*E**VS**V*K*NL***M*DI**LKFS********ML******
```

Fig. 8. Alignment of the predicted VP2 amino acid sequences of six serotypes of BTV. Residues homologous to BTV-10 are indicated by *asterisks* and conserved cysteines are marked by a *dot*.
Variable regions are *inderlined*

```
BTV-10   MEEFVIPVFS ERDIPYSLLN HYPLAIQIDV KVDDEGGKHN LIKIPESDMI DVPRLSIIEA LNYRPKRNDG   70
BTV-11   ********** *TE******S *****VRTN* *IANVDEG*D VV******** ****V**V** *AAK*T****
BTV-17   ********Y* *DE***A**S R******TN* *IEDVE**** VV******** *I*K*T*V** M**K*A****

BTV-10   VVVPRLLDIT LRAYDNRKSA KNAKGVEFMT DTKWMKWAID DKMDIQPLKV TLDNHCSVNH QLFNCIVKAR  140
BTV-11   I********* *****D**AM *S*R****** NA******** *R******** AI*D*NA*** **********
BTV-17   I********* *****D***T *S*R*I**** NAR******* *R******** ***HY***** *****V***N

BTV-10   SANADTIYYD YYPLENGAKR CNHTNLDLLR SLTTTEMFHI LQGAAYALKT TYELVAHSER ENMSESYQVG  210
BTV-11   P*****V**S *F**RDKV*K ********** G********M ******C**S S***ITN*** N*TE*T*AP*
BTV-17   A********* *F***DYK** ********** ***NM*L**A ******SI*S S*****Y*** GSLE*T*V**

BTV-10   TQRWIQLRKG TKIGYRGQPY ERFISSLVQV IIKGKIPDEI RTEIAELNRI KDEWKNAAYD RTEIRALELC  280
BTV-11   VHNR*R*VR* *R***K*EA* S**V****** R*Q*QT*P** VDD**R**E* RT**I**QF* S*K*******
BTV-17   QPK**H*TR* *R**NS*LS* ******M*** SVN******* AN***Q**** RA**IT*T** *GR*******

BTV-10   KILSAIGRKM LDVQEEPKDE MALSTRFQFK LDEKFIRTDQ EHVNIFKVGG SATDDGRFYA LIAIAGTDTQ  350
BTV-11   ********** *NTH****** *D******** **D**KK**S **I***N**A P**HE***** *****A****
BTV-17   S***T***** *NTH****** *D******** *****N*A*S ******G*R* P***E***** *****A****

BTV-10   QGRVWRTNPY PCLRGALIAA ECELGDVYFT LRQTYKWSLR PEYGQRERPL EDNKYVFARL NLFDTNLAVG  420
BTV-11   R********* ********** **Q*****H* ***V****** QD**RT*V** *N*****S*I ****S**E**
BTV-17   K********* *******V** ********S* **RV*T**** *****H*RQ* *N*****N*I ****S*****

BTV-10   DEIIHWRYEV YQPKETTHDD GYICVSQKGD DELLCEVDED RYKEMFDRMI QGGWDQERFK LHNILTEPNL  490
BTV-11   *QVV**K**I DG*A***Y*N ****KTERE* G**V*KIS*E K**T*L**** ********** *YSV**D***
BTV-17   *Q******** KASA***Y*S **M*RHEAEE *****KIN** K****L**** ********** ******D***

BTV-10   LTIDFEKDAY LGARSELVFP PYYDKWINSP MFNARLKIAR GEIATWKADD PWSNRAVHGY IKTSAESLEY  560
BTV-11   ********** *NI***F*L* S*F*Q**Y** ******R*TH ***G*R*SA* **NK*V*F** V*A*T**P**
BTV-17   ********** *NS*****L* D*F****S** ******R*TK ***G*S*K** **N****R** **PL****DF

BTV-10   ALGPYYDLRL QLFGDTLSLG QRQSAVFEHM AQQDDFSTLT DYTKGRTVCP HSGGTFYTFR KVALIILSNY  630
BTV-11   ***Q*F*T*I **Y**A***K *S*****Q*Q S**E**PV** S*A**DV*** ****AL**** ****MLMA**
BTV-17   V********* LF*DE****K *E*****QYL S*L***PA** -QLR*DA*** ****AL**** ****FLIG**

BTV-10   ERLDPSLHEG REHETYMHPA VNDVFRRHVL EMKDFSQLIC FVFDYIFEKH VQLRNAKEAR RIIYLIQNTS  700
BTV-11   ***S*D**** M*DH**T**S IGGANQKRI* **R******* *II*****R* D***DMR*** **L**V*SLG
BTV-17   *K*S*D**** M**QR*V**S T**TYQKRV* ****SC**T* **I******R E***DT**** Y*V****SLT

BTV-10   GAYRLDVLRA EFPNFLKHVM NLRDVKRICD LNVINFFPLL FLVQDNISYW HRQWSIPMIL FDQVIRLIPV  770
BTV-11   EPQ*****SV AS***SRYFL K*K**Q**S* ******L*** **I******* ****AV**** Y*DT*K****
BTV-17   *TQ**S***S T****FQRLL M*KEI*FVR* ******L**M ***H****** ********V* **DT*K****

BTV-10   EVGAYANRFG LKSFFNFIRF HPGDSKKRQD ADDTHKEFGS ICFEYYTTTK ISQGEIDVPV VTSKLDTLKL  840
BTV-11   ********** I******T** ****A****K *********L *S*N**AN** *A**GVHT** **T******I
BTV-17   ********** F***M**T** ***EL**K*I *E*I*****V VA*****N** ****NVHT** M*T*M*V*RV

BTV-10   HVASLCAGLA DSLVYTLPVA HPKKSIVLII VGDDKLEPQI RSEQIVNKYY YSRRHISGVV SICVNQGGQL  910
BTV-11   *LS******* **V******* ****C***** ********HV ****V*S*** F**K*V**** ***IG*ND**
BTV-17   *LS******* **V******* ****C***** ********HT ******SR*N ***K**C*I* *VTIG*NS**

BTV-10   KVHSMGITRH RICDKSILKY KCKVVLVRMP GHVFGNDELM TKLLNV                            956
BTV-11   **Y*S**V** ***E*F**R* *******K** *Y******** ******
BTV-17   R**TS**VK* *V***F***H ****I***** *Y******** ******
```

Fig. 9. Amino acid sequence alignments of VP2 proteins of three BTV serotypes using the BTV-10 as a reference. The conserved amino acids are indicated by *stars*

viral proteins, while at the same time changing the nature of the portions located on the surface of the BTV particle due to immunological pressure from its host.

2.6.3 Segment 3, VP3

This is a major structural protein of BTV inner capsids and it contains group-specific antigenic determinants (Huismans and Erasmus 1981). The full sequence

BTV 13	MEELVIPVIT	ERFDKRLVGR	YDYVIELARP	EGDEWSGHDV	THIPDRRMFD	50
BTV 2	*D**G**IYG	RNYPEH*LKG	*EFL*NTGVK	YPSQGGR***	SK**EMFAY*	
BTV 13	IKVQPIRDAI	DYKPVENDGE	VLPRILDMSI	ACYDMRKRMM	KRDGVDFV-S	100
BTV 2	**DEG****L	K*M*TR***V	*****V*I*L	KG**I**SVI	EANKSNSFHT	
BTV 13	DTKWLEWMIQ	DSMDVQPLKV	DMKEDHSTVQ	YDMFSAKLHV	DSRKADTTSY	150
BTV 2	**S*VQ***K	****Q****I	SID*E**R*V	HSL*NCQVKI	*AK****L**	
BTV 13	NILALETKEG	AQCHHVHTNI	WNHMIRNHLF	HAVQESCYIF	KPTYKLTVNS	199
BTV 2	HLE*I*DA*K	*-*L*TRGQL	C**LT*MD*L	**A**IA*AI	****Q*I*H*	
BTV 13	ERRTPDEDFQ	IGNPQFLTLR	RNQQIFLGDD	AYKKTAKGLV	QVLVNGVVPD	249
BTV 2	**ASTSDN*E	L*RQDVI***	*GHRVQM**E	**T*LMER**	RLT*Q*N**R	
BTV 13	IIRNEIAALD	AIRDKWIQGN	YERTHIKSLE	LCNLLSAIGR	KMVNLEEEPK	299
BTV 2	K*QS**EQ*E	***TR*AT*R	*DPA**N*QD	**RI**R***	I*LDQ*A**V	
BTV 13	DERDLSLRFQ	HKLDDKFAKN	DQERNVIFAQ	KSQRNDQDRF	YVLMVIAASD	349
BTV 2	**DS******	RA**E**RL*	*S***K**EH	**HKK*E***	***LR*****	
BTV 13	TNNSRVWWSN	PYPCLRGALI	AAECKLGDVY	YKLRSWYEWS	VREGYKPRDL	399
BTV 2	*Y***I****	*******T**	***T******	FT*****D**	**SSYIP**R	
BTV 13	DRQYEKYIVG	RVNLFDLEAE	PGTKVLHWEY	ELISKLYTVS	NHEGNQCDLH	449
BTV 2	E*ET****FS	KI****Y**G	*SS**I****	Q*YKRERV*T	LER**P***Y	
BTV 13	PDEGE--IVT	KFDDTRYSDM	IQTIINEGWK	QNDFKMFKML	KDEGNPLLYD	499
BTV 2	***DDEV*I*	****AK**E*	VGE**DG**N	DEE***Y*L*	QEK**V*TI*	
BTV 13	LEKDIKLDRV	SRVVFPPYFD	QWTYVPMFNA	RIKPCEVEVG	ERKNIDPYVK	549
BTV 2	F***T**YNT	*E**L*D*YG	K*IVA****S	KMRII*T*IA	TN*SD**MI*	
BTV 13	RTHRPLKADC	IELMRYHMSQ	YMDLRVSLQG	TSLSIKQTPS	SIHQSLARDA	599
BTV 2	**LK*MTD*P	V**Q**TLAR	*Y*I*PG*M*	R**NRT**Q*	TFDAKVSELP	
BTV 13	SYAEILSRRR	ENLDYKSQCP	IVTNLFLLEK	FFLLIFTTME	KHYWEMDDDE	649
BTV 2	D*EKVV**FG	VIKKPTRP*V	TL*GRYI***	YS**LIDILK	Y*TEVEGNPQ	
BTV 13	TEYEHPKIDP	SKFEVEG-TL	HDVSQVMVHL	FDRFFEKRRF	LRTVDESRWI	698
BTV 2	E*FT**R***	-Q*KFN*N**	S*LN*TV*FI	V*YLH***NY	V*SIY*A*Y*	
BTV 13	LHLIRSASGA	RRLEVLSRFF	PAFSDGLRI-	REFKKVRDIM	LLNFLPFLFL	748
BTV 2	ISR***ST**	A*MSIIEFY*	*T*ARLISNA	**PTY*K*L*	A*****L**I	
BTV 13	TGDNIAYEHR	QWAVPVIFYA	DKIMIIPAEV	GAYYNRFGLT	CILELMMFFP	798
BTV 2	V***MI*K**	**SI*LLL*T	*RVKV**L**	*SSN**Q*FV	SY**Y*F***	
BTV 13	SYDTRNENLS	EDVRACIGPI	INYYLDTTIS	NGGIQTSIVS	TKALLYETYL	848
BTV 2	*LAD*TSKVD	*SMIKVSKEM	V***MK****	E**VNLNV**	**S***DI**	
BTV 13	SSICGGFSEA	ILWYLPITHP	SKCLIALEVS	DALTSPELRI	DKIKRRFPLS	898
BTV 2	**V***V*DG	VV********	Y**VV*I**C	*DRVPAR**C	*RL*L*****	
BTV 13	SNHLKGIVQI	SVRPGRTFSV	VTQGIVKHRV	CKKTLLRYRC	DVILIQTPGY	948
BTV 2	******I***	QINEEGG*D*	Y*E***T***	***S**KHV*	*IV**KFH*H	
BTV 13	VFGNDELLTK	LLNI				
BTV 2	******M***	***V				

Fig. 10. Amino acid sequence alignments of VP2 proteins of BTV-13 and BTV-2. The conserved amino acids are indicated by *stars*

BTV 2	MDELGIPIYG	RNYPEHLLKG	YEFLINTGVK	YPSQGGRHDV	SKIPEMFAYD	50
BTV 1	*********K	*GF*****H*	***T*DSST*	IQ*V******	T*L***N***	
BTV 2	IKDEGIRDAL	KYMPTRNDGV	VLPRIVDISL	KGYDIRKSVI	EANKSNSFHT	100
BTV 1	**S**M*T**	W*N*V****F	****VL**T*	R***GKRA**	DSSRHK****	
BTV 2	DTSWVQWMIK	DSMDQQPLKI	SIDEEHSRVV	HSLFNCQVKI	DAKKADTLSY	150
BTV 1	*ER*****M*	****A****V	GL*DQTQKIA	***H**V***	*S*****M**	
BTV 2	HLEAIEDAEK	A-CLHTRGQL	CNHLTRMDLL	HAAQEIAYAI	KPTYQLIVHS	199
BTV 1	*I*P***SL*	G-*****TM*	W***V*VEMS	*********L	****DIV**A	
BTV 2	ERASTSDNFE	LGRQDVITLR	RGHRVQMGDE	AYTKLMERLV	RLTVQGNVPR	249
BTV 1	**RDR*QP*R	P*D*TL*NFS	**QK***NHN	S*E*MV*G*A	H*VIR*KT*E	
BTV 2	KIQSEIEQLE	AIRTRWATGR	YDPAHINSQD	LCRILSRIGR	IMLDQEAEPV	299
BTV 1	L*RD**TK*D	E*CN**IRS*	***GE*KAYE	**KV**TV**	K*****K**A	
BTV 2	DEDSLSLRFQ	RALDEKFRLN	DSERNKIFEH	KSHKKDEDRF	YVLLRIAASD	349
BTV 1	**AN**I***	E*I*N***QH	****L*****	RNQRR*****	*I**M*****	
BTV 2	TYNSRIWWSN	PYPCLRGTLI	AAETKLGDVY	FTLRSWYDWS	VRSSYIPRER	399
BTV 1	*F*T*V****	**********	*S********	SMM*L*****	**PT***Y*K	
BTV 2	ERETEKYIFS	KINLFDYEAG	PSSKVIHWEY	QLYKRERVVT	LERGNPCDLY	449
BTV 1	S**Q****YG	RV*****V*E	*GT*I***A*	K*NQQIKDI*	Y*Q******F	
BTV 2	PDEDDEVIIT	KFDDAKYSEM	VGEIIDGGWN	DEEFKMYKLL	QEKGNVLTID	499
BTV 1	**_***A*V*	****VA*GQ*	*NDL*N***D	Q*R***H*I*	KSQ*******	
BTV 2	FEKDTKLYNT	SEVVLPDYYG	KWIVAPMFNS	KMRIIETEIA	TNKSDDPMIK	549
BTV 1	****A**TSN	EG*AM*E*FD	***I*****A	*L**KHG**T	QRRN****V*	
BTV 2	RTLKPMTDDP	VELQRYTLAR	YYDIRPGLMG	RSLNRTQTQS	TFDAKVSELP	599
BTV 1	***S*IAFA*	IV***L****	F*****AI**	QA*S*Q*G**	*Y*EEI*KIE	
BTV 2	DYEKVVSRFG	VIKKPTRPCV	TLTGRYILEK	YSLLLIDILK	YHTEVEGNPQ	649
BTV 1	G*AEILQ*R*	IVQI*KK**P	*V*AQ*T**R	*A*F**N**E	Q*IIQSTDED	
BTV 2	EEFTHPRIDP	-QFKFNGNTL	SDLNQTVVFI	VDYLHEKRNY	VRSIYEARYI	699
BTV 1	VMYS***V*Y	KLEVH*ENI	I*IS*I*I*V	F*F*F*R*RT	**GV**S**M	
BTV 2	ISRIRSSTGA	ARMSIIEFYF	PTFARLISNA	REPTYVKDLM	ALNFLPLLFI	749
BTV 1	VT***DAQ*Q	N*INV*TEF*	***GYHL*RV	K*A*IIQEI*	Y******F*L	
BTV 2	VGDNMIYKHR	QWSIPLLLYT	DRVKVIPLEV	GSSNNRQGFV	SYLEYMFFFP	798
BTV 1	*S**I**T*K	***V**F**A	HEL*******	**Y*D*CSL*	**I***V***	
BTV 2	SLADRTSKVD	ESMIKVSKEM	VNYYMKTTIS	EGGVNLNVVS	TKSLLYDIYL	848
BTV 1	*K*F****L*	*VQP*IAR*M	LK**IN*K*F	***I*****T	**Q***ET**	
BTV 2	SSVCGGVSDG	VVWYLPITHP	YKCVVAIEVC	DDRVPARLRC	DRLKLRFPLS	898
BTV 1	A*I***L***	I*********	N**L*****S	*E****SI*A	SHI*******	
BTV 2	AQHLKGIVVI	QINEEGGFDV	YTEGIVTHRV	CKKSLLKHVC	DIVLLKFHGH	948
BTV 1	VK******I*	*VD***K*T*	*S****S***	***N***YM*	*******S**	
BTV 2	VFGNDEMLTK	LLNV				
BTV 1	**********	****				

Fig. 11. Homology between VP2 protein of US serotypes 2 and Australian serotype 1. The conserved amino acids are indicated by *stars*

Table 4. Homology among six BTV VP2 proteins

	BTV-10	BTV11	BTV-13	BTV-17	BTV-2
BTV-11	70.2				
BTV-13	32.0	39.0			
BTV-17	71.4	72.5	38.7		
BTV-2	39.9	39.9	47.2	41.1	
BTV-1	39.8	39.7	45.8	40.9	58.6

Percent of each VP2 amino acid sequence homology between the serotypes is scored using a computer alignment program

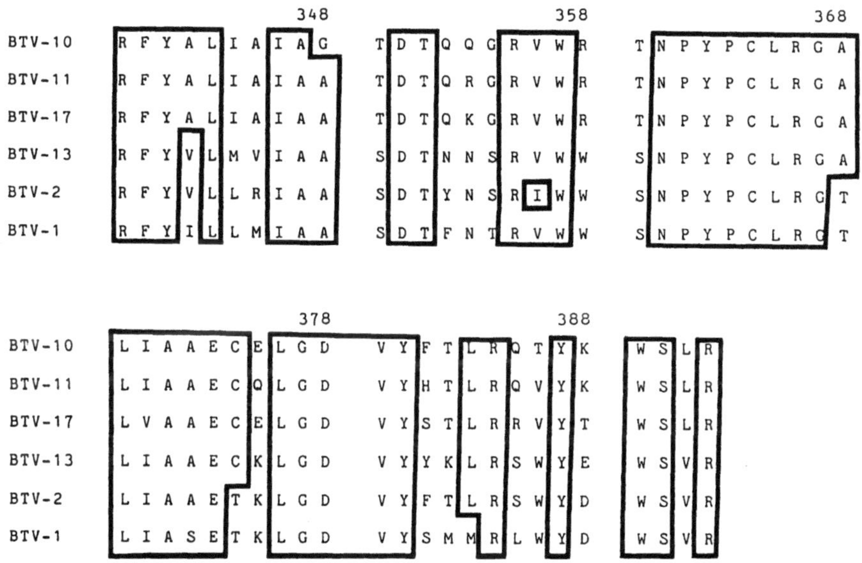

Fig. 12. Conserved amino acid sequences of six VP2 proteins. The single mutation from isoleucine (I) to valine (V) is *circled*

of this segment is known for BTV-10, BTV-17, and BTV-1 (Australian isolate) (GHIASI et al. 1985; PURDY et al. 1984; GOULD 1987), and partial sequences are known for the South African isolates of BTV-1 and BTV-9, and the Australian isolates of BTV-9 and BTV-15 (GOULD 1987). The sequence is highly conserved among serotypes. For example, there are 126 point differences between the nucleotide sequences of BTV-10 and BTV-17, and this corresponds to only nine changes at the amino acid level which is 0.15% of the possible sites where a single nucleotide change would cause an altered amino acid. By contrast, 114 of the nucleotide changes are silent (i.e., 6.5% of all possible silent changes). A similar pattern is seen upon comparing the VP3 sequence of BTV-1 with those of BTV-10 and BTV-17 (GOULD 1987). The amino acid composition is virtually identical for all three serotypes (GHIASI et al. 1985; GOULD 1987) with a low content of charged amino acids and a high content of hydrophobic amino acids (Fig. 7).

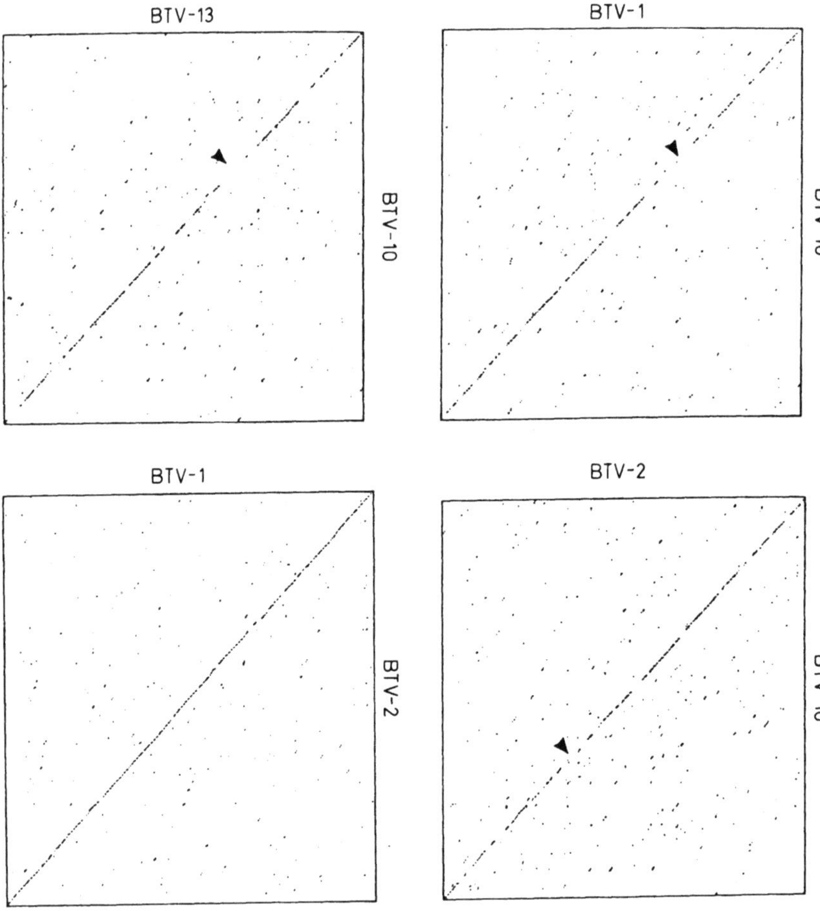

Fig. 13. Diagon analyses. The complete amino acid sequences of predicted L2 gene products of BTV-10, BTV-1 or -2 or -13 and BTV-2 and -13 were compared pairwise using the Diagon program of STADEN (1982). Homologies are indicated by *diagonal lines*. For the protein comparison an 11-amino acid window and a proportional index of 131 was employed

2.6.4 Segment 4, VP4

This protein is a minor component of the inner core and its function is not known. The 5′ noncoding region of this segment is the shortest of all the segments at only 8 bp (Table 1); the functional implications of this are not known. Currently the sequence of this segment is only known for BTV-10 (YU et al. 1987). Hybridization studies indicate that the sequence is well conserved among serotypes. VP4 of BTV-10 has a high content of charged amino acids (particularly histidine) and tryptophan but a low content of glutamine. The protein is more hydrophilic in nature, with a strong hydrophilic domain in the carboxyl terminus (Fig. 7).

```
                    10        20        30        40        50        60        70        80        90       100
BTV 10      MGKIIKSLSRFGKKVGNALTSNTAKKIYSTIGKAAERFAESEIGAATIDGLVQGSVHSIITGESYGESVKQAVLLNVLGTGEELPDPLSPGERGMQTKIK
BTV 2       ***V*R**********S*****A*********************S*AT***********L***********************I********Q*****L*
BTV 1 AUS   ***V*R**************************************S*A************L*******Y***************I*****Q**TEI*A*LR
BTV 1 SA    ***V*R**N**************************DE*L*****S*AF***********************************I**********I*A*L*

                   110       120       130       140       150       160       170       180       190       200
BTV 10      ELEDEQRNELVRLKYNKEITEKFGKELGEVYDFMNGEAKEVEAVEEQYTMLCKAVDSYEKILKEEDSKMAILARALQREAAERSEDEIKMVKEYRQKIDA
BTV 2       ***E************DK*K****E**EQ**E****A***EVGA*K*FDI*S***N**N***T***QQ*RR**I***K*IG**THV*TA******N****
BTV 1 AUS   ****************DK*K****E**E***E****A**AEVED*K*FDI*N***T**N***T***LQ*RR**N***K*IG**THA*TV******N****
BTV 1 SA    ****************DK*K***E***E***N*Y****NAEIED*K*FDI*N***T**N***T***LQNRR**TR**K*IG**THA*TV******D****

                   210       220       230       240       250       260       270       280       290       300
BTV 10      LKSAIEIERDGMQEEAIQEIAGMTADVLEAASEEVPLIGAGMATAVATGRAIEGAYKLKKVINALSGIDLSHMRSPKIEPTIIATTLEHRFKDIPDEQLA
BTV 2       ************************************************************************L*T*****SVVS*I**Y*A*****NA**
BTV 1 AUS   **N*********************RG**********************************************L*T*****SVVS*I**Y*TRA***SA**
BTV 1 SA    **N***V*****************************************************************FTHYENR*SVVS*I**Y*A*E***NA**

                   310       320       330       340       350       360       370       380       390       400
BTV 10      ISVLNKKTAVADNCNEIAHIKQEILPKFKQIMNEEKEIEGIEDKVIHPRVMMRFKIPRTQQPQIHIYAAPWDSDDVFFFHCVSYHHRNESFFLGFDLGID
BTV 2       V***S*NR*IQE*HK*LM***D****R**KA*D*****C*****T***K***K*****S********T*************I*H**A*****I****S**
BTV 1 AUS   V***S*NR*IQE*HK*LI***D****R**KA*D*****C*****T***K*********A******V*S*************I*H**A**********S**
BTV 1 SA    V***I*ER*IQE*HK*LM***N****R**KA*D*****MRDRRQM***K***K*****A******V*S*************I*H**A*D********S**
                                                                                     ●
                   410       420       430       440       450       460       470       480       490       500
BTV 10      VVHFEDLTSHWHALGMAQEASGRTLTEAYREFLNLSISSTFSSAIHARRMIRSRAVHPIFLGSMHYDITYEALKNNAQRIVYDDELQMHILRGPLHFQRR
BTV 2       L**Y****A******A**M*V********K***K*A****YGTQM*T**L***KT****Y***L****SFAD*RG*************************
BTV 1 AUS   ***Y****A******A**M*M****S***K****M***NAYGTQM*T**LV**KM****Y***L****SFLD*RG*******************I*****
BTV 1 SA    L**Y****A*A**I*A**T*A**************A**KA*GTQM*T**LV**KT****Y***L****SFSD*RG*******************I*****

                   510            526
BTV 10      AILGALKFGVKILGDKIDVPLFLRNA
BTV 2       *********C*****RL********P
BTV 1 AUS   *********C*V***RL*********
BTV 1 SA    *********C*V***RL*********
```

Fig. 14. Alignment of the predicted VP5 amino acid sequences of four isolates of BTV. Residues homologous to BTV-10 are indicated by *asterisks* and the position of the conserved cysteine is marked by a *dot*

2.6.5 Segment 5, VP5

This protein, along with VP2, forms the outer capsid of bluetongue virions. Hybridization studies (see above) show that this RNA segment is not highly conserved among serotypes. This segment has been sequenced for BTV-10 (PURDY et al. 1986), both Australian and South African isolates of BTV-1 (GOULD and PRITCHARD 1988; WADE-EVANS et al. 1988), and BTV-2 (our unpublished data). The overall homology between these protein sequences (Fig. 14) is 70%. However, when conservative changes are taken into account, it is 82%. As in VP2 of BTV-1 and BTV-2, VP5 of US serotype 2 appears to be more closely related to serotype 1 of both Australian and South African isolates in comparison with that of US serotype 10.

The changes are distributed throughout the protein and it is difficult to define variable regions as one can do for VP2 (see Sect. 2.6.2). Only one of the cysteine residues (at position 381) is conserved among all four sequences. Overall predicted charge, at neutral pH, varies being + 4.5, + 1.5, and + 1.0 for BTV-10 and BTV-2, the Australian and South African isolates of BTV-1, respectively. The protein is rich in certain nonpolar amino acids, such as alanine and isoleucine, and has a low content of tryptophan. There are at least three strong hydrophobic regions, including the carboxyl terminus (Fig. 7; PURDY et al. 1986; FUKUSHO et al. 1989).

2.6.6 Segment 6, NS1

This segment is highly conserved among serotypes (see above) and codes for a nonstructural protein, NS1, which forms tubules in the cytoplasm of BTV-infected cells (HUISMANS 1979; HUISMANS and ELS 1979) and, using *Spodoptera frugiperda*, cells infected with a recombinant baculovirus (URAKAWA and ROY 1988). The protein is rich in cysteine, tryptophan, and tyrosine and has the lowest content of serine and threonine, per unit length, of all the BTV proteins (LEE and ROY 1987). It also has a low content of charged amino acids. This segment has been sequenced for the Australian and South African isolates of BTV-1 (GOULD et al. 1988), and when these are compared with the sequence from BTV-10 there is an overall homology of 87%, at the amino acid level, which rises to 96% when conservative changes are taken into account. All 16 of the cysteine residues, in the BTV-10 sequence, are conserved in the three sequences, which suggests that NS1 has a highly ordered, disulfide bond-linked, structure. The protein also has several regions of hydrophobic amino acids throughout the molecule, particularly in the carboxyl terminal half (Fig. 7). The role that these features play in the formation of NS1 secondary and quaternary (i.e., tubule) structure has yet to be determined.

2.6.7 Segment 7, VP7

This protein is the major structural component of BTV cores and comprises 36% of the total core protein (HUISMANS et al. 1987a). It contains group-specific antigenic determinants (Huismans and Erasmus 1981; Gumm and Newman 1982). Only the sequence of BTV-10 segment 7 has been reported to date (Yu et al. 1988). The amino acid composition of VP7 differs considerably from that of the other BTV gene products. It has a very low proportion of charged amino acids (i.e., R + K + $\frac{1}{2}$H = 80; D + E = 77) as compared with the average for BTV gene products of 142 and 134 respectively (Table 2). The proportion of asparagine and glutamine, however, at 109 per 1000 residues (Table 2) is much higher than the average (i.e., 77). VP7 has only one lysine which is remarkable since the other gene products are lysine rich. The protein is richer in alanine, methionine, and proline than are the other BTV gene products, and it is particularly rich in hydrophobic sequences (Fig. 7; Yu et al. 1988).

There is a discrepancy between the predicted size of VP7 (38 548 daltons) and its estimated size by SDS-PAGE (29 kD) (Huismans 1979) which may be caused by its aberrant mobility due to its hydrophobic nature. It has been postulated that VP7 is located on the surface of the inner core, but Hyatt and Eaton (1988) have recently reported that VP7 may be accessible from the surface of the virion. In this context it is noteworthy that segment 7 is not as well conserved among serotypes as are the segments coding for the other inner core proteins (Huismans and Cloete 1987; Squire et al. 1985; Ritter and Roy 1988).

2.6.8 Segment 8, NS2

This is a nonstructural protein and it is the only BTV encoded protein that is phosphorylated in BTV-infected cells (Huismans et al. 1987b; Devaney et al. 1988). The protein has the lowest proportional content of threonine of all the BTV gene products but an average content of serines. NS2 has a high content of cysteines, mostly at the C-terminal end, which suggests a highly ordered structure (Fukusho et al. 1989). Overall the protein is hydrophilic (Fig. 7), rich in charged amino acids, but with a low number of histidines. The sequence of segment 8 has also been reported for a South African isolate of BTV-10 (Hall et al. 1989) and it has 95% homology, at the amino acid level, with that of BTV-10 from the USA. NS2 from BTV-infected BHK cells has been shown to be capable of binding BTV single-stranded RNA (Huismans et al. 1987b). Both termini of the protein contain hydrophobic sequences and the secondary structure prediction for the sequence is rich in β-turns.

2.6.9 Segment 9, VP6

This protein is a minor component of BTV inner cores and is hydrophilic (Fig. 7) with a unique amino acid composition by comparison with the other BTV gene products (Fukusho et al. 1989). The protein has only one cysteine and a low proportional content of aromatic, asparagine, and glutamine residues. VP6 is very rich in glycine (40 out of 328 amino acids, 12%) with 12 between residues 88 to 116, including five consecutive glycines (Fukusho et al. 1989). VP6 is the richest of all the BTV gene products in charged amino acids and it is highly conserved among serotypes (see above). Although the function of VP6 is not known, our preliminary studies indicate that the protein is associated with the viral RNA in BTV inner cores (Roy et al. 1989).

2.6.10 Segment 10, NS3

The complete nucleotide sequence of the segment has been reported for BTV-10 (Lee and Roy 1986) and for the Australian BTV-1 (Gould 1988b). This nonstructural protein has a predicted mass of 0.5×10^6 Daltons which is somewhat higher than the estimate made from PAGE by Verwoerd et al. (1970). Both the known sequences contain two initiation codons in the same, long, open reading frame (i.e., one starting at position 20 and the other at position 59). The predicted size of the translated products from these two initiation codons is 25 572 and 24 020 Daltons, respectively. In vitro translation of segment 10 RNA yields two proteins of approximately these sizes in equimolar amounts (Sangar and Mertens 1983; Mertens et al. 1984; Van Dijk and Huismans 1988).

The segment 10 sequences of the two serotypes are highly homologous (82%) with both termini being particularly well conserved (e.g., two changes in the first 66 nucleotides at the 5′ end). The reason for the 3′ noncoding region being three to four times longer (at 113 bp) than the other segments is not known. NS3 is rich in serine and threonine residues but deficient in arginine, tryptophan, glycine, and isoleucine by comparison with the other BTV gene products. There are strong hydrophobic regions towards the carboxyl end of the molecule (Fig. 7, Lee and Roy 1986). To date no function has been assigned to NS3.

In summary, analyses of the oligonucleotide fingerprints, hybridization studies, and complete cDNA sequences have confirmed that all the genes of BTV genome segments representing the nonstructural proteins (NS1, NS2, and NS3), as well as most of the inner core proteins (VP1, VP3, VP4, VP6, and VP7), are well conserved among serotypes. In contrast, the segments coding for the two outer capsid proteins, particularly VP2, vary considerably. However, in spite of these variations in amino acid sequence, secondary structure predictions suggest that the overall three-dimensional protein structures of VP2 and VP5 are well conserved. Despite geographical separation, it is apparent that all BTV serotypes have a common ancestor but have evolved differently, some as closely related groups and others independently.

2.7 Expression of BTV Genome Segments in Insect Cells Using Recombinant Baculoviruses

The availability of complete cDNA clones representing the ten discrete dsRNA segments and the availability of their sequence database have been further exploited for understanding the structural/functional relationships of the BTV proteins. Each of the cDNA clones has been manipulated and expressed to a high level in insect cells using recombinant baculoviruses. Studies with these recombinant viruses and the expressed proteins have only begun to contribute to our understanding of the structural/functional relationships of various BTV gene products. In addition, by manipulating the vector systems, we have embarked on defining the architecture of the complex morphology of the viral capsids. The description of those BTV proteins and morphological structures which have proved of particular interest will be discussed below following a brief summary of the technology of baculovirus expression.

2.7.1 The Baculovirus Expression System

This expression system utilizes the major late promoter of the polyhedrin gene in *Autographa californica* nuclear polyhedrosis virus (AcNPV) (reviewed by MILLER 1988). The life cycle of this virus is characterized by the production of two forms of progeny; extracellular virus particles (ECVs) and occluded virus particles (OVs). ECVs are produced relatively early in infection (from 12 h onwards) and are released by budding from the cell surface. They mediate the systemic infection of the insect and also account for the mode of infection in cell culture. Later in the infection cycle (from 18 h onwards) viral progeny are occluded into a paracrystalline matrix composed primarily of a 29 kDa protein called polyhedrin. These occlusions, called polyhedra, protect the progeny virus during horizontal transmission and effect their release in a new host by dissolving in the alkaline environment of the insect gut. Polyhedra accumulate in infected cells for 4–5 days until cell lysis, by which time polyhedrin may constitute up to 50% of the total cell protein. Because cell-to-cell infection is propagated by ECVs, the synthesis of polyhedrin protein is a nonessential function for the replication of AcNPV in cell culture. The use of AcNPV as an expression system, therefore, involves replacement of the polyhedrin gene with a foreign gene which, due to the control of the polyhedrin promoter, has the potential to be expressed to a high level.

A number of transfer vectors are available to construct such recombinant baculoviruses. One which has found particular favor is pAcYM1 (MATSUURA et al. 1987). This, in common with most other baculovirus expression vectors, consists of a restriction enzyme fragment of the AcNPV genome encompassing the polyhedrin gene, cloned into a high copy-number bacterial plasmid. The polyhedrin gene sequence has been deleted in pAcYM1 and replaced by a *Bam*H1 linker to allow for the insertion of a foreign gene immediately downstream of the polyhedrin promoter. The unchanged wild-type AcNPV DNA sequences that flank the inserted gene mediate homologous recombination when a cell is transfected with the plasmid DNA and wild-type AcNPV DNA. Recombinant baculoviruses are selected on the basis of their polyhedrin-negative phenotype (see Fig. 15).

By means of this in vivo recombination technique, cDNA copies of all ten BTV segments (principally from serotype 10) have been expressed in the baculovirus system (see Fig. 16). The first point of note is that aside from segment 10, which encodes two closely related polypeptides, each BTV segment has been confirmed to be monocistronic. The synthesis of individual BTV polypeptides in

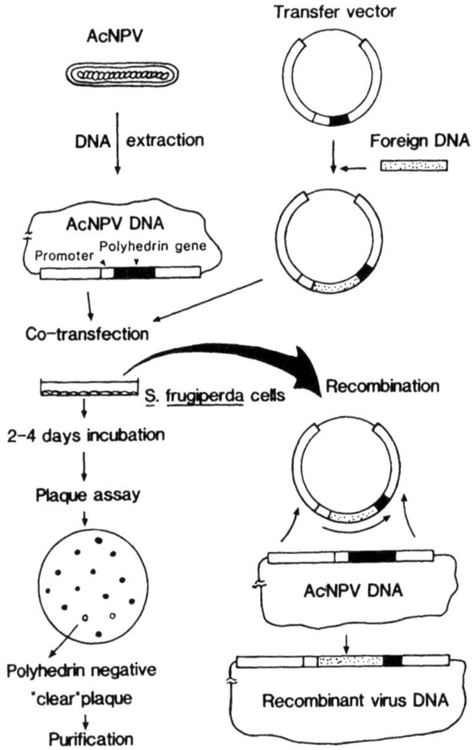

Fig. 15. General protocol for the insertion of foreign DNA into infectious *Autographa californica* nuclear polyhedrosis virus: Foreign DNA inserted at a locus of a cloned AcNPV subgenomic fragment is introduced into *Spodoptera frugiperda* culture together with infectious AcNPV DNA by transfection procedures. In vivo recombination occurs between the AcNPV DNA sequences flanking the foreign insert and homologous DNA sequences on the replicating AcNPV genome forming a novel recombinant DNA molecule which can in turn be packaged into infectious recombinant AcNPV virus within 48-72 h

cell culture, both structural and nonstructural, enables a variety of questions to be addressed. Some of the more significant contributions to our understanding of BTV morphology as well as the function of individual proteins resulting from analysis of expressed protein are described below.

2.7.2 VP2 and VP5: The Outer Capsid Proteins

VP2, one of the major outer capsid proteins, is the antigen recognized by neutralizing antibodies (Huismans and Erasmus 1981). VP2 may be a suitable candidate for a subunit vaccine since Huismans et al. (1987c) demonstrated that VP2 purified from whole virus particles can elicit neutralizing antibodies in sheep and protect against challenge with the same serotype. Although the success of such trials is a promising development in the quest for a BTV subunit vaccine, the relatively large quantity of purified virus required to prepare a single protective dose makes it impractical to scale-up this method for commercial production. The synthesis of VP2 using genetic engineering techniques therefore provides an alternative strategy.

Inumaru and Roy (1987) have shown that VP2 of serotype 10 expressed in insect cells by a recombinant baculovirus is authentic in terms of its antigenic

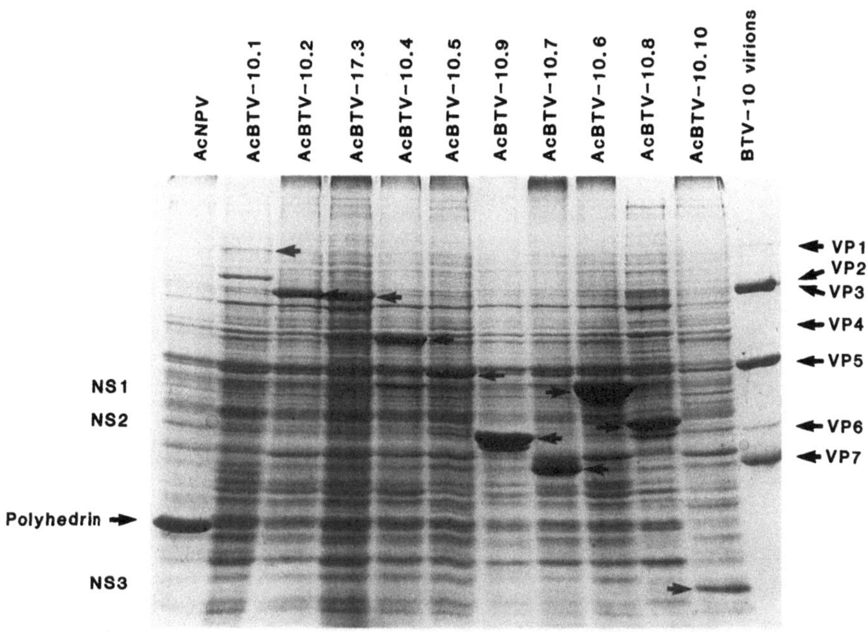

Fig. 16. Expression of BTV proteins by recombinant baculoviruses. *S. frugiperda* cells were infected at a multiplicity of 10 pfu/cell with wild-type AcNPV or recombinant baculovirus containing a BTV gene. Cells were harvested at 48 h postinfection and after separation by SDS-PAGE proteins were located by Coomassie Brilliant Blue R staining

properties. Furthermore, antisera raised in mice or rabbits following immunization with crude preparations of this expressed protein neutralized the infectivity of not only BTV-10 virus but also, to a lesser extent, BTV-11 and BTV-17 viruses. The segment 2 genes of serotypes 1, 2, 10, 11, 13, and 17 have now been expressed in the baculovirus system (see Fig. 17), and work is currently in progress to raise antisera to these proteins and test their cross-neutralizing activity among homologous and heterologous serotypes.

A further aspect of virus neutralization which has been substantiated by use of the baculovirus expression system is the inability of VP5 alone to induce antibodies that neutralize BTV virus (Table 5, Marshall and Roy 1989). Huismans et al. (1987c) have previously reported that a mixture of solubilized VP2 and VP5 elicits a higher titer of neutralizing antibodies than solubilized VP2 alone. To assess the role of VP5 individually in virus neutralization, a recombinant baculovirus containing the *M5* gene of BTV-10 was constructed. Unlike other AcNPV-BTV recombinant viruses, this construct had a toxic effect on insect cells. Electron micrographs of infected cells showed a loss of cytoplasmic and nuclear material and of increase in size of vesicular bodies, which may have been derived from autophagic lysosomes in cells infected with the recombinant virus. Due to the reduction in cell viability, the expression level of VP5 was considerably lower in comparison with the other BTV proteins that

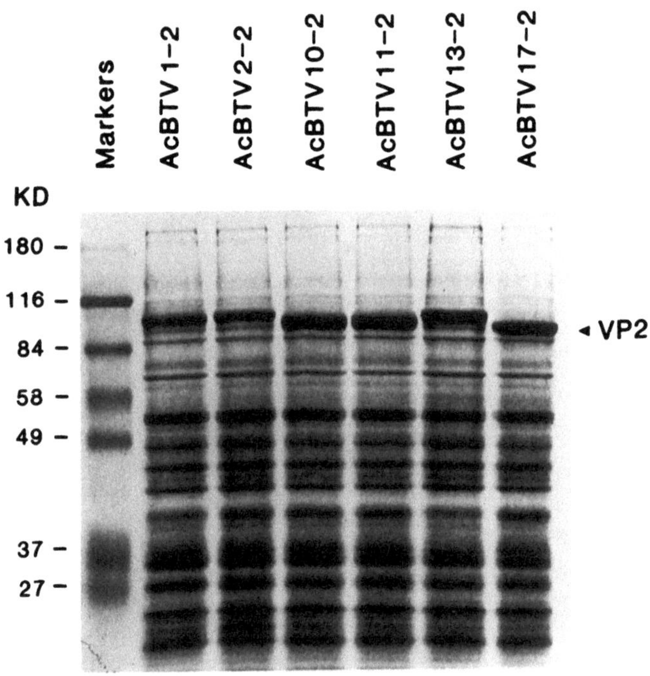

Fig. 17. Expression of VP2 proteins from six BTV serotypes by recombinant baculoviruses. Samples were prepared as in Fig. 15 and proteins stained with Coomassie Brilliant Blue R

Table 5. Plaque reduction neutralization titers of antisera raised to expressed VP2 and VP5

Antisera	BTV serotypes			
	10	11	13	17
Rabbit VP2 antisera	> 640	> 160	0	> 160
Preimmune rabbit sera	0	0	0	0
Mouse VP5 ascitic fluid	0	0	0	0
Control ascitic fluid	0	0	0	0
Mouse antisera to:				
YM1/10–2 infected *S. frugiperda* cells	205 ± 74* ($n = 4$)	—	—	—
YM1/10–5 infected *S. frugiperda* cells	51 ± 23⁺ ($n = 4$)	—	—	—
AcNPV infected *S. frugiperda* cells	55 + 40 ($n = 4$)	—	—	—

*Significantly different from AcNPV infected *S. frugiperda* cells at the $P = 0.05$ level
⁺ Not significantly different from AcNPV infected *S. frugiperda* cells at the $P = 0.05$ level

have been expressed by this system. Nevertheless, the expressed VP5 was recognized by antisera raised to BTV-10 virus and the antisera raised to recombinant VP5 protein recognized authentic BTV-10 VP5 protein. However, unlike VP2 proteins, the anti-VP5 antibodies failed to neutralize the infectivity of

BTV-10 in vitro (MARSHALL and ROY 1989) (Table 5). VP5, therefore, does not appear to play a direct role in virus neutralization, although a synergistic interaction with VP2 cannot be ruled out.

To test the ability of these recombinant proteins to protect against virus infection, sheep were inoculated with whole cell extracts of insect cells infected with recombinant baculoviruses. Unlike the mice or rabbits, the group of sheep which received a mixture of VP2 and VP5 accumulated higher titers of neutralizing antibodies than the group that had received only VP2 alone. However, both groups were totally protected against challenge with virulent virus (unpublished observations). These results indicated that although large quantities ($> 100\,\mu g$) of VP2 alone are adequate for protection, together with VP5 the protective capabilities are probably much more efficient. Further studies on the interactions between these two proteins and their role in viral infection and possible subunit vaccines are in progress.

2.7.3 Core Protein Morphology

Little is known about the functions, interactions, or stoichiometry of the five core proteins. Appreciation of such factors is obviously important if we are to understand the morphology of a complex virus such as BTV. Recent cryoelectron microscopy studies suggest that the core of BTV consists of a nucleoprotein center surrounded by two distinct protein layers, each of which is composed of a single polypeptide species (our unpublished observations). Immunogold analysis indicates that VP7 is the principal component of the outermost layer and is attached to a framework of VP3 (HYATT and EATON 1988). We were therefore interested to see if we could synthesize BTV core particles from these components, and to this end the segment 3 and segment 7 genes were inserted into a baculovirus dual expression transfer vector. Such vectors incorporate two copies of the polyhedrin promoter and transcription termination sequences (EMERY and BISHOP 1987) with a unique enzyme restriction site located downstream of each promoter to allow for the insertion of two foreign genes. The promoters are present in opposite orientations to minimize the possibility of homologous sequence recombination and excision of one or other of the foreign genes. Recombinant baculoviruses were prepared by the established procedure of cotransfecting *S. frugiperda* cells with the dual expression plasmid DNA and wild-type AcNPV DNA (FRENCH and ROY 1990).

Electron micrographs of *S. frugiperda* cells infected with the recombinant baculovirus showed large aggregates of foreign material in the cytoplasm which, under high magnification, appeared to consist of spherical particles. This expressed material was isolated by lysing the cells with NP40 and purification on a discontinuous sucrose gradient. When examined under the electron microscope the material was found to consist of empty core-like particles whose size and appearance were indistinguishable from authentic BTV core particles prepared from BTV-infected BHK cells (Fig. 18).

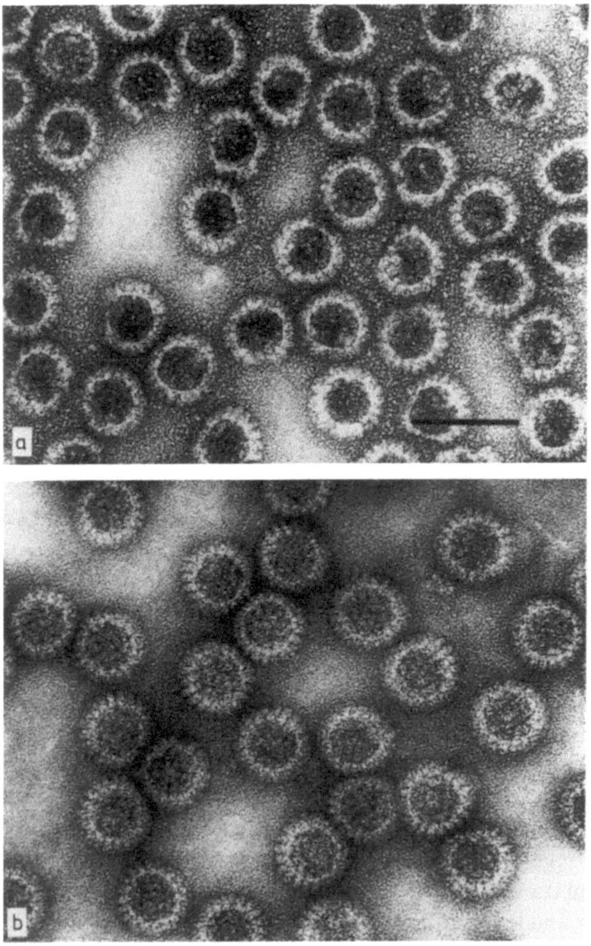

Fig. 18 a. Electron micrographs of empty BTV core particles synthesized in insect cells by a recombinant baculovirus expressing the major BTV core proteins VP3 and VP7. **b** Authentic BTV core particles prepared from BTV-infected BHK-cells are included for comparison

How precisely the expressed particles resembled their viral counterparts was further appraised by calculation of the VP3 to VP7 stoichiometry. It has been estimated to be from 15 VP7 molecules to 2 VP3 molecules in the expressed particle. This matches exactly the stoichiometry of these components in BTV core particles as determined by Verwoerd et al. (1972). Recent evidence from our laboratory indicates that 780 copies of VP7 are present per core particle which, based upon a 15:2 molecular arrangement, suggests that the cores contain 104 copies of VP3.

The synthesis of empty BTV core particles from the major core proteins VP3 and VP7 reveals some important aspects of BTV morphology. It can be concluded that their formation is not dependent on the presence of the three minor core proteins, nor the BTV dsRNA. In addition, the three BTV nonstructural proteins (NS1, NS2, NS3) are not required to assist or direct the assembly of the VP3 and VP7 proteins to form stoichiometrically correct empty particles.

In this study the *L3* segment gene was isolated from serotype 17, while the segment 7 gene originated from serotype 10. VP3 is very highly conserved among serotypes and sequence data have shown there to be a 99% homology at the amino acid level between serotypes 10 and 17 (GHIASI et al. 1985). The potential of singularly expressed VP3 from serotype 17 to act as an antigen for diagnosis has been examined by indirect ELISA. It was confirmed that VP3 synthesized in insect cells by a recombinant baculovirus reacted with sheep antisera raised to BTV-1, BTV-2, BTV-10, BTV-11, BTV-13, and BTV-17 (INUMARU et al. 1987; our unpublished data).

Purified BTV core particles have been shown to contain an RNA-directed RNA polymerase (VERWOERD and HUISMANS 1972). VP1 has been considered a candidate for the virion replicase/transcriptase based upon its size, location, molar ratio in the core, and predicted amino acid sequence (ROY et al. 1988). To test whether VP1 is indeed the virion replicase/transcriptase, *S. frugiperda* cells were infected with a recombinant baculovirus expressing the VP1 protein and a cell lysate containing solubilized VP1 tested for polymerase activity. It was evident from these experiments that significant levels of radioactivity were

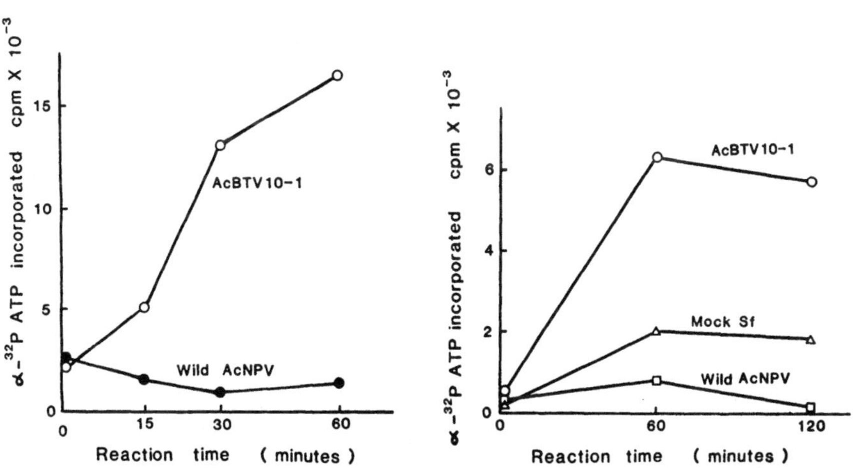

Fig. 19 a,b. Polymerase activity of a recombinant baculovirus containing VP1 protein of BTV-10: In a standard reaction, a 25-μl aliquot of supernatant recovered from infected cell lysates was assayed at 37°C in 200 μl of reaction mixture containing polyu (20 μg/ml), oligo (A) 12–18 (10 μg/ml), and ^{32}P-ATP. The enzyme activity was measured by the incorporation of radioactivity in the TCA precipitable material. **A** Recombinant virus infected lysate (o), wild-type virus lysate (●). **B** Recombinant virus in lysate (o), wild-type virus lysate (□), and mock-infected cell lysate (△)

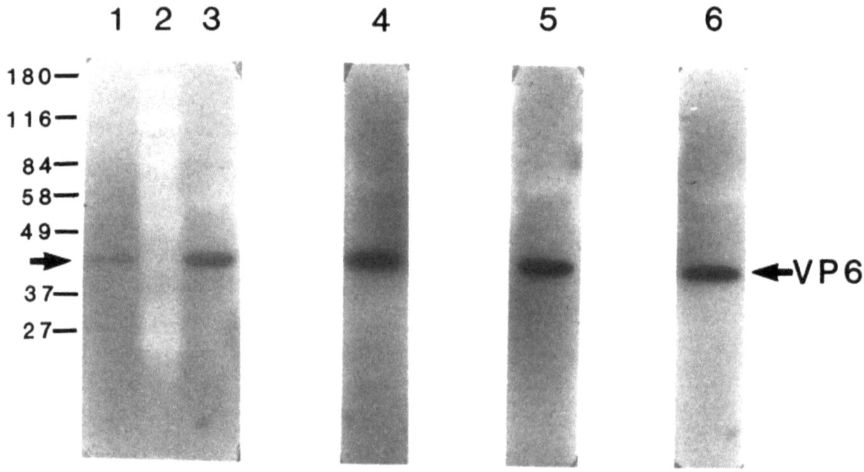

Fig. 20. Proteins were analysed by SDS-PAGE and transferred to an Immobilon filter. The protein blot was cut into strips and individual strips were reacted with the (*lane 3*) dsRNA probe (see Materials and Methods). *Lane 4* BTV ssRNA probe obtained by SP6 transcription of the BTV-10 segment 10 cDNA clone. *Lane 5*, dsDNA probe derived from pUC19 and, *lane 6*, ssRNA probe derived by SP6 transcription of the pST-18 plasmid. *Lane 1* contains purified BTV-10 virion proteins and *lane 2* represents the marker proteins mixture

recovered only from reaction mixtures in which the cell lysate originated from recombinant baculovirus infected cells (Fig. 19; Urakawa et'al. 1989), and not from mock or AcNPV-infected cells. It was therefore concluded that the incorporation of labelled ATP resulted from the newly synthesized oligonucleotide chains by the active recombinant BTV polymerase (i.e., VP1).

VP6, which is also a minor core protein, is rich in charged amino acids. This suggests that it may function as an RNA-binding protein. In an RNA overlay protein blot assay VP6, from both baculovirus-expressed and from purified virions, binds both ds and ss BTV RNA, indicating that it may be closely associated with the viral genome and is a component of the viral nucleocapsid (Fig. 20; Roy et al. 1990). The characteristics and specificity of this binding and its functional role in virus assembly are under investigation.

2.7.4 Virus-Coded Nonstrucutural Proteins

Two virus-specific entities, tubules and granular inclusion bodies, are routinely observed in BTV-infected cells (Lecatsas 1968). The tubules appear to be composed of only one type of protein, NS1 (Huismans and Els 1979), which is the gene product of the *M6* segment. To obtain direct proof that NS1 forms tubules, the *M6* gene was expressed in the baculovirus system. When insect cells were infected with the derived recombinant baculovirus, tubular structures similar to those reported in BTV-infected cells were made (see Fig. 21), and following purification were shown to consist of the NS1 protein (Urakawa and Roy 1988).

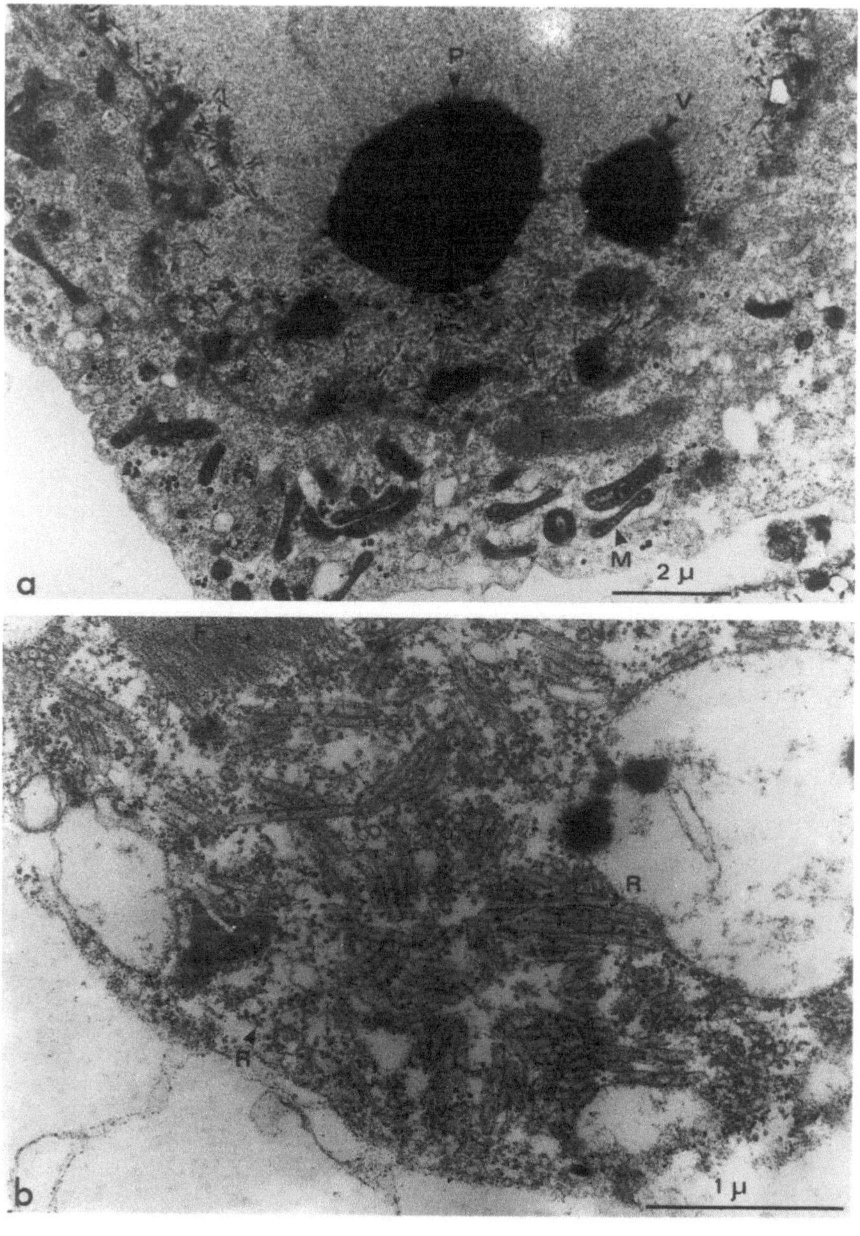

Fig. 21a, b. (*Continued*)

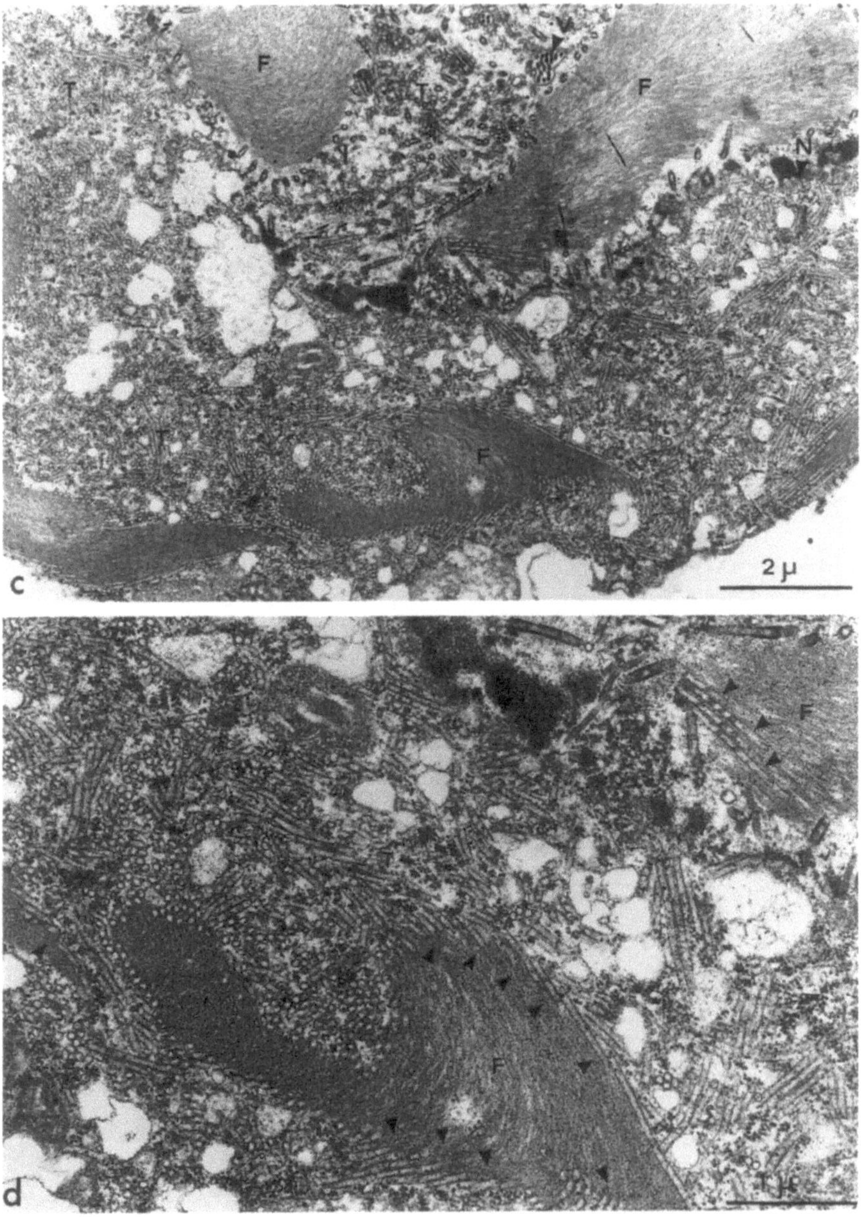

Fig. 21 a–d. Electron micrograph of baculovirus-expressed tubules. *S. frugiperda* cells infected with **a** AcNPV or, **b–d** a recombinant virus were fixed with 2% glutaraldehyde 72 h postinfection and processed for electron microscopy. *P*, Polyhedrin; *V*, virus particles; *M*, mitochondria; *R*, ribosomes; *F*, fibrous structure; *T*, tubules; *N*, nuclear membrane

Another aspect of NS1 morphology which was examined following its expression was its potential to act as an antigen for BTV diagnosis. Antisera raised in response to a BTV infection are known to contain antibodies to NS1 (HUISMANS and ELS 1979). The *M6* gene of BTV-10 is highly conserved among 20 BTV serotypes and baculovirus-expressed NS1 reacted strongly in an ELISA test with polyclonal sheep antisera raised to BTV-2, BTV-10, BTV-11, BTV-13, or BTV-17, but not with normal sheep sera. NS1 would thus seem to be a suitable candidate for a diagnostic reagent, particularly since there appears to be limited homology with the corresponding segments of two EHD virus serotypes (as shown by hybridization studies).

NS2 protein, which is the gene product of RNA segment S8 and the only virion-directed phosphoprotein, has been expressed in high levels (see Fig. 16). The expressed protein has also been demonstrated not only to be essentially similar in terms of its size, peptide maps profile, and antigenicity to the authentic BTV NS2 protein, but also phosphorylated similarly (as shown in Fig. 22). Moreover, the expressed protein has been shown to bind ssRNA species. Using the gold-labelled anti-NS2 monospecific polyclonal antisera in an immuno-electron microscopic study, NS2 protein has been localized within the virus inclusion bodies but not with the virus particles in the BTV-infected BHK-21 cells (THOMAS and ROY, manuscript in preparation).

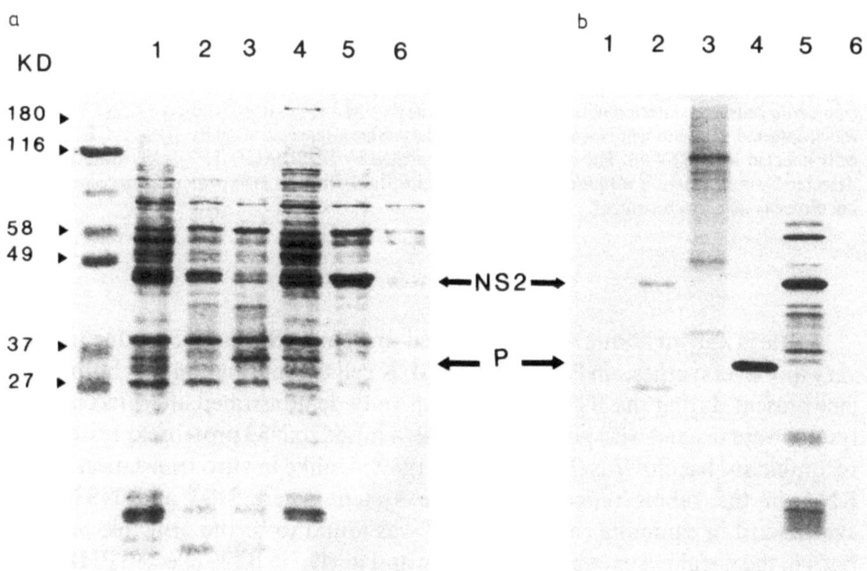

Fig. 22 a,b. Expression of NS2 by recombinant baculovirus analysed using PAGE **a** Coomassie Blue stained gel. **b** Autoradiogram of the gel: *1*, ^{32}P-labelled AcNPV infected cell lysate; *2*, ^{32}P-labelled recombinant virus infected cell lysate; *3*, ^{32}P-labelled mock infected cell lysate; *4–6*, tracks similar to 1-3 except labelled with ^{35}S-labelled cell lysates

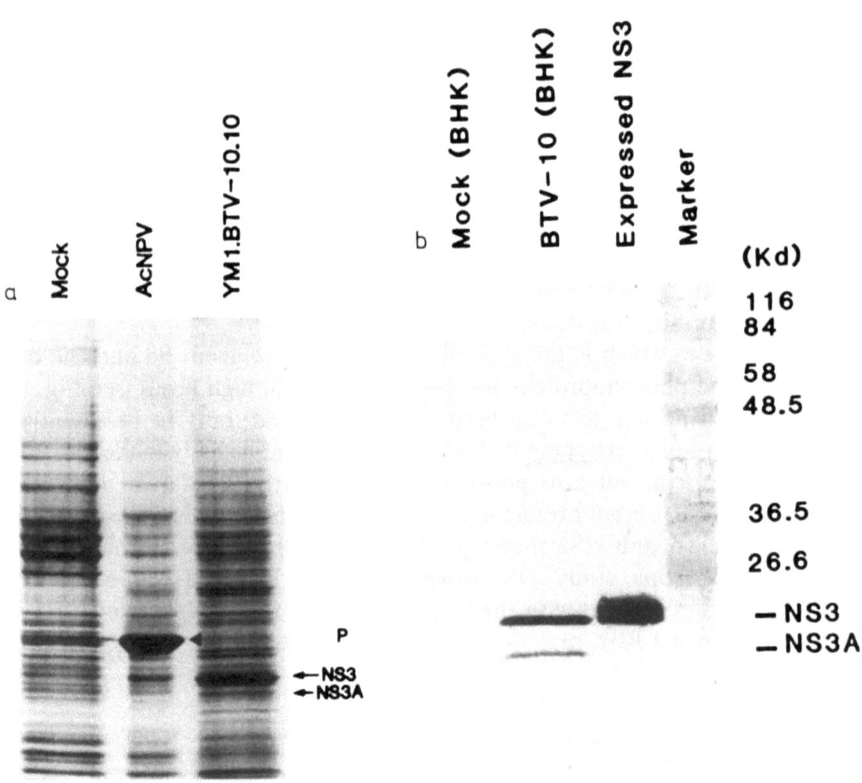

Fig. 23a, b. Expression of NS3 by recombinant baculovirus and Western blot analysis. **a** *S. frugiperda* cells were infected with recombinant virus AcYM 1 BTV-10.10 wild-type AcNPV or were mock-infected. Proteins were located using Coomassie Blue after separation by SDS-PAGE. **b** BHK cells infected with BTV-10. The proteins were separated by SDS-PAGE (15%), electroblotted, and detected by their reaction with mouse anti-NS3 ascitic fluid. Purified NS3 protein from recombinant baculovirus acted as a control

Little is known about the S10 encoded proteins NS3 and NS3A due to their very low level synthesis in BTV-infected BHK cells. Confirmation that both are in fact present during the BTV life cycle was only demonstrated after infected cell lysates were reacted with polyclonal antisera raised to NS3 protein expressed by a recombinant baculovirus (French et al. 1989). Unlike in vitro translation of S10 RNA in the rabbit reticulocyte lysate system where NS3 and NS3A are synthesized in equimolar amounts, NS3 was found to be the principle product both in the baculovirus expression system and in vivo in BTV-infected BHK cells (Fig. 23). Such a result indicates the caution which should be exercised when using the in vitro rabbit reticulocyte lysate system to predict the pattern of protein synthesis from a gene with alternative start codons. It was also found that NS3 and NS3A expressed in insect cells reacted strongly with sera from sheep infected with homologous and heterologous serotypes suggesting that the S10 products are highly conserved group-specific antigens.

3 Conclusions

The importance of the genetic relationships among BTV serotypes has been appreciated for many years, yet until recently the genetic variation and relatedness of BTV serotypes have been difficult to analyse in detail. The recent development in recombinant DNA and nucleic acid sequencing technology has enabled rapid accumulation of not only the complete sequences of the whole genome of one BTV serotype, but also sequence data on a number of different genes of other TV serotypes. This information has led to new insights into the evolution of the viruses and accurate characterization of various genes. In addition, the predicted sequence analyses of various viral-coded proteins have allowed the assessment of the structural/functional relationships of the various genes and gene products.

By the use of novel baculovirus single and multiple expression vectors, we have just embarked on defining the role of various proteins in virus replication, multiplication cycles, and various morphological structures, which undoubtedly will eventually unravel the various processes of protein-protein interaction in the morphogenetic events of the virus.

References

Borden EC, Shope RE, Murphy FA (1971) Physicochemical and morphological relationships of some arthropod-borne viruses to bluetongue virus–a new taxonomic group. Physiochemical and serological studies. J Gen Virol 13: 261–271

Bowne JG, Ritchie EA (1970) Some morphological features of bluetongue virus. Virology 40: 903–911

Collisson EW, Roy P (1983) Analyses of the genome of bluetongue virus vaccines and a recent BTV isolate of Washington State. Amer J Vet Res 244: 235–237

Collisson EW, Barber TL, Gibbs PJ, Greiner EC (1985) Two electropherotypes of bluetongue virus serotype 2 from naturally infected calves. J Gen Virol 66: 1279–1286

Cowley JA, Gorman BM (1987) Genetic reassortments for identification of the genome segment coding for the bluetongue virus hemagglutinin. J Virol 61(7): 2304–2306

Cowley JA, Gorman BM (1989) Cross-neutralization of genetic reassortants of bluetongue virus serotypes 20 and 21. Vet Microbiol 19: 37–51

Devaney MA, Kendall J, Grubman MJ (1988) Characterization of a nonstructural phosphoprotein of two orbiviruses. Virus Res 11: 51–164

Els HJ (1973) Electron microscopy of bluetongue virus RNA. Onderstepoort J Vet Res 40: 73

Els HJ, Verwoerd DW (1969) Morphology of bluetongue virus. Virology 38: 213–219

Emery VC, Bishop DHL (1987) The development of multiple expression vectors for high level synthesis of AcNPV polyhedrin protein by a recombinant baculovirus. Protein Engineering 1: 359–366

French TJ, Roy P (1990) Synthesis of empty BTV core particles by a recombinant baculovirus expressing the two major structural core proteins. J Virol (in press) 64
French TJ, Inumaru S, Roy P (1989) Expression of two related non-structural proteins of BTV-10 in insect cells by a recombinant baculovirus: production of polyclonal ascitic fluid and characterization of the gene product in BTV-infected BHK cells. J Virol 63: 3270–3278
Fukusho A, Ritter GD, Roy P (1987) Variation in the bluetongue virus neutralization protein VP2. J Gen Virol 68: 2967–2973
Fukusho A, Yu Y, Yamaguchi Y, Roy P (1989) Completion of the sequence of bluetongue virus serotype 10 by the characterization of structural protein VP6 and a non-structural protein, NS2. J Gen Virol 70: 1677–1689
Ghiasi H, Purdy MA, Roy P (1985) The complete sequence of bluetongue virus serotype 10 segment 3 and its predicted VP3 polypeptide compared with those of BTV serotype 17. Virus Res 3: 181–190
Ghiasi H, Fukusho A, Eshita Y, Roy P (1987) Identification and characterisation of conserved and variable regions in the neutralization VP2 gene of bluetongue virus. Virology 160: 100–109
Gorman BM, Taylor J, Walker PJ, Davidson WL, Brown F (1981) Comparison of bluetongue type 20 with certain viruses of the bluetongue and eubenangee serological groups of orbiviruses. J Gen Virol 57: 251–261
Gorman BM, Taylor J, Walker PJ (1983) Orbiviruses. In: Joklik WK (ed) The reoviridae. Plenum, New York, pp 287–357
Gould AR (1987) The complete nucleotide sequence of bluetongue virus serotype 1 RNA 3 and a comparison with other geographic serotypes from Australia, South Africa and the United States of America, and with other orbivirus isolates. Virus Res 7: 169–183
Gould AR (1988a) Conserved and non-conserved regions of the outer coat protein, VP2, of the Australian bluetongue serotype 1 virus revealed by sequence comparison to the VP2 of North American BTV serotype 10. Virus Res 9: 145–158
Gould AR (1988b) Nucleotide sequence of the Australian bluetongue virus serotype 1 RNA segment 10. J Gen Virol 69: 945–949
Gould AR, Pritchard LI (1988) The complete nucleotide sequence of the outer coat protein, VP5, of the Australian bluetongue serotype 1 reveals conserved and non-conserved sequences. Virus Res 9: 285–292
Gould AR, Pritchard LI, Tavaria MD (1988) Nucleotide and deduced amino acid sequences of the non-structural protein, NS1, of Australian and South African bluetongue serotype 1. Virus Res 11: 97–107
Grubman MJ, Appleton JA, Letchworth GJ (1983) Identification of bluetongue virus type 17 genome segments coding for polypeptides associated with virus neutralization and intergroup reactivity. Virology 131: 355–366
Gumm ID, Newman JFE (1982) The preparation of purified bluetongue virus group antigen for use as a diagnostic reagent. Arch Virol 72: 83–93
Hall SJ, Van Dijk AA, Huismans H (1989) Complete nucleotide sequence of gene segment encoding non-structural protein NS2 of SA bluetongue virus serotype 10. Nucleic Acids Res 17: 457
Howell PG (1960) A preliminary antigenic classification of strains of bluetongue virus. Onderstepoort J Vet Res 28: 357–363
Howell PG (1970) The antigenic classification and distribution of naturally occurring strains of bluetongue virus. J S Afr Vet Med Assoc 41: 215–223
Howell PG, Verwoerd DW (1971) Bluetongue virus. In: Hess WR, Howell PG, Verwoerd DW (eds) African swine fever virus, bluetongue virus. Springer, Berlin Heidelberg New York, pp 35–74 (Virology monographs, vol 9)
Huismans H (1979) Protein synthesis in bluetongue virus-infected cells. Virology 92: 385–396
Huismans H, Howell PG (1973) Molecular hybridization studies on the relationships between different serotypes of bluetongue virus and on the difference between virulent and attenuated strains of the same serotype. Onderstepoort J Vet Res 40: 93–104
Huismans H, Els HJ (1979) Characterization of the microtubules associated with the replication of three different orbiviruses. Virology 92: 397–405
Huismans H, Bremer CW (1981) A comparison of an Australian bluetongue virus isolate (CSIRO 19) with other bluetongue virus serotypes by cross-hybridization and cross-immune precipitation. Onderstepoort J Vet Res 48: 59–67
Huismans H, Erasmus BJ (1981) Identification of the serotype-specific and group-specific antigens of bluetongue virus. Onderstepoort J Vet Res 48: 51–58
Huismans H, Cloete M (1987) A comparison of different cloned bluetongue virus genome segments as probes for the detection of virus-specified RNA. Virology 158: 373–380

Huismans H, Van Dijk AA, Els HJ (1987a) Uncoating of parental bluetongue virus to core and subcore particles in infected L cells. Virology 157: 180–188
Huismans H, Van Dijk AA, Bauskin AR (1987b) In vitro phosphorylation and purification of a nonstructural protein of bluetongue virus with affinity for single-stranded RNA. J Virol 61: 3589–3595
Huismans H, Van Der Walt NT, Cloete M, Erasmus BJ (1987c) Isolation of a capsid protein of bluetongue virus that induces a protective immune response in sheep. Virology 157: 172–179
Hyatt AD, Eaton BT (1988) Ultrastructural distribution of the major capsid proteins within bluetongue virus and infected cells. J Gen Virol 69: 805–815
Inumaru S, Roy P (1987) Production and characterization of the neutralization antigen VP2 of bluetongue virus serotype 10 using a baculovirus expression vector. Virology 157: 472–479
Inumaru S, Ghiasi H, Roy P (1987) Expression of bluetongue virus group specific antigen VP3 in insect cells by a baculovirus expression vector: its use for detection of bluetongue virus antibodies. J Gen Virol 68: 1627–1637
Kahlon J, Sugiyama K, Roy P (1983) molecular basis of bluetongue virus neutralization. J Virol 48: 627–632
Knudson DL, Shope RE (1985) Overview of the orbiviruses. In: Barber TL, Jochim MM (eds) Bluetongue and related orbiviruses. Liss, New York, pp 255–266
Kowalik TF, Li JKK (1987) The genetic relatedness of United States prototype bluetongue viruses by RNA/RNA hybridization. Virology 158: 276–284
Kozak M (1981) Possible role of flanking nucleotides in recognition of the AUG initiator codon by eukaryotic ribosomes. Nucleic Acids Res 9: 5233–5252
Lecatsas G (1968) Electron microscopic study of the formation of bluetongue virus. Onderstepoort J Vet Res 35: 139–149
Lee J, Roy P (1986) Nucleotide sequence of a cDNA clone of RNA segment 10 of bluetongue virus (serotype 10). J Gen Virol 67: 2833–2837
Lee J, Roy P (1987) Complete sequence of the NS1 gene (M6 RNA) of US bluetongue virus serotype 10. Nucleic Acids Res 15: 7207
Marshall JJA, Roy P (1990) High level expression of the two outer capsid proteins of bluetongue virus serotype 10; their relationship with the neutralization of virus infection. Virus Res (in press)
Matsuura Y, Possee RD, Overton HA, Bishop DHL (1987) Baculovirus expression vectors: the requirements for high level expression of proteins, including glycoproteins. J Gen Virol 68: 1233–1250
Mertens PPC, Sangar DV (1985) Analysis of the terminal sequences of the genome sequences of four orbiviruses. Virology 140: 55–67
Mertens PPC, Browns F, Sangar DV (1984) Assignment of the genome segments of bluetongue virus type 1 to the proteins they encode. Virology 135: 207–217
Mertens PPC, Pedley S, Cowley J, Burroughs JN (1987) A comparison of six different bluetongue virus isolates by cross-hybridization of the dsRNA genome segments. Virology 161: 438–447
Mertens PPC, Pedley S, Cowley J, Borroughs JN, Corteyn AH, Jeggo MH, Jennings DM, Gorman BM (1989) Analysis of the roles of bluetongue virus outer capsid proteins VP2 and VP5 in determination of virus serotype. Virology 170: 561–565
Miller LK (1988) Baculoviruses as gene expression vectors. Annu Rev Microbiol 42: 177–199
Murphy FA, Borden EC, Shope RE, Harrison A (1971) Physiochemical and morphological relationships of some arthropod-borne viruses to bluetongue virus—a new taxonomic group: electron microscopic studies. J Gen Virol 13: 273
Oberst RD, Squire KRE, Stott JL, Chuang R, Osburn BI (1985) The coexistence of multiple bluetongue virus electropherotypes in individual cattle during natural infection. J Gen Virol 66: 1901–1909
Osburn BE, McGowan B, Heron B, Loomis E, Bushnell R, Stott J, Utterback W (1981) Epizootiologic study of bluetongue: virologic and serologic results. Am J Vet Res 42: 884–887
Owen NC, Munz EK (1966) Observations on a strain of bluetongue virus by electron microscopy. Onderstepoort J Vet Res 33: 9–14
Pedley S, Mohamed MEH, Mertens PPC (1988) Analysis of genome segments from six different isolates of bluetongue virus using RNA-RNA hybridisation: a generalised coding assignment for bluetongue viruses. Virus Res 10: 381–390
Purdy MA, Petre J, Roy P (1984) Cloning of the bluetongue virus L3 gene. J Virol 51: 754–759
Purdy MA, Ghiasi H, Rao CD, Roy P (1985) Complete sequence of bluetongue virus L2 RNA that codes for the antigen recognized by neutralizing antibodies. J Virol 55: 826–839

Purdy MA, Ritter GD, Roy P (1986) Nucleotide sequence of cDNA clones encoding the outer capsid protein, VP5, of bluetongue virus serotype 10. J Gen Virol 67: 957–962
Rao CD, Roy P (1983) Genetic variation of bluetongue virus serotype 11 isolated from hosts (sheep) and vectors (*Culicoides variipennis*) at the same site. Am J Vet Res 44: 911–914
Rao CD, Sugiyama K, Roy P (1983a) Evolution of bluetongue virus serotype 17. Am J Trop Med Hyg 32: 865–870
Rao CD, Kiuchi A, Roy P (1983b) Homologous terminal sequences of the genome double-stranded RNAs of bluetongue virus. J Virol 46: 378–383
Ritter DG, Roy P (1988) Genetic relationships of bluetongue virus serotypes isolated from different parts of the world. Virus Res 11: 33–47
Roy P, Sugiyama K, Collisson E, Rao CD, Kahlon J (1982) Isolation of recombinant bluetongue virus and their genetic analysis. In: Mackenzie JS (ed) Viral diseases in Southeast Asia and the Western Pacific Academic, pp 637–641
Roy P, Ritter GD, Akashi H, Collisson E, Inaba Y (1985) A genetic probe for identifying bluetongue virus infections in vivo and in vitro. J Gen Virol 66: 1613–1619
Roy P, Fukusho A, Ritter DG, Lyons D (1988) Evidence for genetic relationship between RNA and DNA viruses from the sequence homology of a putative polymerase gene of bluetongue virus with that of vaccinia virus: conservation of RNA polymerase genes from diverse species. Nucleic Acids Res 16: 11759–11767
Roy P, Adachi A, Urakawa T, Booth TF, Thomas CP (1990) Identification of bluetongue virus VP6 protein as a nucleic acid-binding protein and the localization of VP6 in virus-infected vertebrate cells. J Virol 64: 1–8
Samal SK, Livingston CW, McConnell S, Ramig RF (1987) Analysis of mixed infection of sheep with bluetongue virus serotypes 10 and 17: evidence for genetic reassortment in the vertebrate host. J Virol 61: 1086–1091
Sangar DV, Mertens PPC (1983) Comparison of type 1 bluetongue virus protein synthesis in vivo and in vitro. In: Compans RW, Bishop DHL (eds) Double-stranded RNA viruses. Elsevier, New York, pp 183–191
Squire KRE, Chuang RY, Chuang LF, Doi RH, Osburn BI (1985) Detecting bluetongue virus RNA in cell culture by dot hybridization with a cloned genetic probe. J Virol Methods 10: 59–68
Staden R (1982) An interactive graphics program for comparing and aligning nucleic and amino acid sequences. Nucleic Acids Res 10: 2951–2961
Stott JL, Osburn BI, Barber TL (1982) Recovery of dual serotypes of bluetongue virus from infected sheep and cattle. Vet Microbiol 7: 197–207
Studdert MJ, Panborn J, Addison RB (1966) Bluetongue virus structure. Virology 29: 509–511
Sugiyama K, Bishop DHL, Roy P (1981) Analyses of the genomes of bluetongue viruses recovered in the United States. I. Oligonucleotide fingerprint studies that indicate the existence of naturally occurring reassortant BTV isolates. Virology 114: 210–217
Sugiyama K, Bishop DHL, Roy P (1982) Analyses of the genomes of bluetongue virus isolates recovered from different states of the United States and at different times. Am J Epidemiol 115: 332–347
Urakawa T, Roy P (1988) Bluetongue virus tubules made in insect cells by recombinant baculovirus: expression of NS1 gene of bluetongue virus serotype 10. J Virol 62: 3919–3927
Urakawa T, Ritter DG, Roy P (1989) Expression of largest RNA segment and synthesis of VP1 protein of bluetongue virus in insect cells by recombinant baculovirus: association of VP1 protein with RNA polymerase activity. Nucleic Acids Res 17: 7395–7401
Van Dijk AA, Huismans H (1988) In vitro transcription and translation of bluetongue virus mRNA. J Gen Virol 69: 573–581
Verwoerd DW (1969) Purification and characterization of bluetongue virus. Virology 38: 203–212
Verwoerd DW, Huismans H (1972) Studies on the in vitro and in vivo transcription of the bluetongue virus genome. Onderstepoort J Vet Res 39: 185–192
Verwoerd DW, Louw H, Oellermann RA (1970) Characterization of bluetongue virus ribonucleic acid. J Virol 5: 1–7
Verwoerd DW, Els HJ, de Villiers EM, Huismans H (1972) Structure of bluetongue virus capsid. J Virol 10: 783–794
Verwoerd DW, Huismans H, Erasmus BJ (1979) Orbiviruses. In: Fraenkel-Conrat H, Wagner RR (eds) Comprehensive virology vol 14. Plenum, New York, pp 285–345
Wade-Evans AM, Pan ZQ, Mertens PPC (1988) Sequence analysis and in vitro expression of a cDNA clone of genome segment 5 from bluetongue virus, serotype 1 from South Africa (VRR 00446). Virus Res 11: 227–240

Yamaguchi S, Fukusho A, Roy P (1988a) Complete sequence of L2 gene of the bluetongue virus Australian serotype 1 (BTV-1). Nucleic Acids Res 16: 2725

Yamaguchi S, Fukusho A, Roy P (1988b) Complete sequence of neutralization protein VP2 of the recent US isolate bluetongue virus serotype 2: its relationship with VP2 species of other US serotypes. Virus Res 11: 49–58

Yu Y, Fukusho A, Roy P (1987) Nucleotide sequence of the VP4 core protein gene (M4 RNA) of US bluetongue virus serotype 10. Nucleic Acids Res 15: 7206

Yu Y, Fukusho A, Ritter DG, Roy P (1988a) Complete nucleotide sequence of the group-reactive antigen VP7 gene of bluetongue virus. Nucleic Acids Res 16: 1620

The Replication of Bluetongue Virus

B. T. EATON, A. D. HYATT, and S. M. BROOKES

1 Introduction 89

2 The Reovirus Replication Cycle 90

3 Bluetongue Virus Genes and Proteins 92

4 Adsorption of Bluetongue Virus to Cells 93

5 Endocytosis 93
5.1 Inhibition of Virus Replication by Modulators of Endosomal and Lysosomal pH 96

6 Amplification of Infecting Virions 96
6.1 Superinfection of Virus-Infected Cells 98

7 Structures Observed in BTV-Infected Cells 100
7.1 Attachment to Intermediate Filaments 101
7.2 Virus Tubules 103
7.3 Virus Inclusion Bodies 104
7.4 Virus and Virus-Like Particles 106

8 Release of Virus from Infected Cells 108

9 Conclusions 113

References 115

1 Introduction

Bluetongue virus (BTV) replicates in the cytoplasm of a wide variety of cell types and infection ultimately leads to cell death. The studies of VERWOERD, HUISMANS and others in the late 1960s and continuing to the present (see Chap. 2, this volume) on the double-stranded, segmented genomic RNA (VERWOERD 1969; VERWOERD et al. 1970), the bishelled nature of the virus particle (VERWOERD et al. 1972), the activation of the virion-bound transcriptase, and the activity of this enzyme in vivo and in vitro (VERWOERD and HUISMANS 1972) indicated that BTV possesses many characteristics in common with reovirus. However, BTV and the similar African horse-sickness and epizootic hemorrhagic disease (EHD) of deer viruses differ from reovirus in several respects. They are smaller, lack a well-defined outer capsid layer, and exhibit greater pH sensitivity. In addition, they are insect transmitted. Such differences led to the grouping of these viruses (VERWOERD 1970) into a genus for which the name "orbivirus" was proposed (BORDEN et al. 1971). BTV is the type species of the *Orbivirus* genus within the Reoviridae family.

Australian Animal Health Laboratory, C.S.I.R.O., P.O. Bag 24, Geelong Victoria 3220, Australia

Current Topics in Microbiology and Immunology, Vol. 162

Biochemical and morphological aspects of the reovirus replication cycle have been extensively studied and reviewed (JOKLIK 1981; ZARBL and MILLWARD 1983; TYLER and FIELDS 1985). Salient features of the reovirus multiplication cycle including the conservative mode of RNA replication are outlined briefly in Sect. 2. There is little doubt that replication of the genome of other double-stranded RNA viruses such as BTV or rotavirus, proceeds via a similar conservative mechanism. However the protein content, structure, and morphology of the inner and outer capsids of BTV differ from their reovirus counterparts. Consequently it is to be expected, and electron microscopic observation confirms, that this leads to the synthesis of intracellular structures which are unique to BTV-infected cells. In contrast to the situation in reovirus-infected cells, unfortunately there is no information on the structure, protein content, or biophysical properties of particles such as those with replicase or transcriptase activities, and little is known of intermediates in BTV particle morphogenesis.

Until recently information on morphological aspects of BTV replication derived in the main from studies done some 20 years ago on electron microscopic analysis of thin sections of BTV-infected cells (BOWNE and JOCHIM 1967; LECATSAS 1968; CROMACK et al. 1971). There are three factors which led us to reexamine BTV replication from a morphological perspective. First, new and refined techniques in transmission and scanning electron microscopy are now available. Second, the use of monoclonal antibodies especially those labeled with colloidal gold, has permitted the intracellular localization of specific virus proteins. Third, information on the structure of BTV and the amino acid sequence of most of the virus structural and nonstructural proteins provide a framework to relate structure and function of viral proteins (see Chap. 3, this volume).

In this chapter we summarize what is currently known about the replication of BTV and draw attention to several aspects which serve to differentiate the replication cycles of BTV and reovirus. In addition, we show that released progeny BTV particles reenter infected cells and their replication contributes significantly to both the kinetics of production and the final yield of infectious BTV virions. We also indicate how immunoelectron microscopy has contributed to our understanding of some of the late steps in virus morphogenesis and the release of progeny virus from the infected cell.

2 The Reovirus Replication Cycle

Analysis of reovirus-infected cells has revealed a strategy for RNA synthesis and replication that is probably used by all viruses containing double-stranded segmented RNA genomes (JOKLIK 1981). For this reason and to provide a basis for comparison with BTV, key features of the reovirus replication cycle and virus morphogenesis pathway are outlined briefly in this section.

Reovirus are taken into the cell by endocytosis (DALES 1973). Following transport to lysosomes the outer coat of the virus is partially disrupted (STURZENBECKER et al. 1987). The subviral particle (SVP) generated contains an active transcriptase and is probably released into the cytoplasm by crossing the lysosomal membrane (TYLER and FIELDS 1985). The parental SVP synthesize initially four and subsequently ten mRNAs (WATANABE et al. 1968) which are capped at the 5′ terminus (FURWICHI and SHATKIN 1977) and have two functions. Not only may they be translated to generate viral proteins, but they also serve as the role source of positive single-stranded RNAs for encapsidation into progeny virions (ZARBL and MILLWARD 1983). Parental virus RNA transcripts are made early in infection (SKUP and MILLWARD 1980).

Little is known of the way in which reovirus particles are assembled. It is clear, however, that morphogenesis is linked to replication of the genome. The morphogenetic process appears to start with particles which comprise both core and outer coat proteins (in proportions different from that found in virions) and ten single-stranded RNAs of positive polarity (MORGAN and ZWEERINK 1975). The single-stranded RNA binding property of the nonstructural protein σ-NS (HUISMANS and JOKLIK 1976) suggests that although it is not itself incorporated into virions (ZWEERINK et al. 1971), it may play a role in the, as yet obscure, process whereby ten individual single-stranded parental RNA transcripts are sequestered into a precursor particle. These single-stranded RNA-containing particles exhibit a replicase activity which synthesizes RNA of negative polarity thereby generating double-stranded RNA. RNA synthesis and addition of viral proteins to the complex probably occur concomitantly (ACS et al. 1971).

A number of uncharacterized alterations in structure and protein content follow the synthesis of double-stranded RNA and lead to the formation of progeny subviral particles with transcriptase activity. These particles contain predominantly core proteins and variable amounts of the nonstructural protein, μ-NS (MORGAN and ZWEERINK 1975; ZWEERINK et al. 1976). If replicase particles are direct precursors of transcriptase particles, μ-NS may be responsible for displacing the outer coat proteins associated with replicase particles (ZARBL and MILLWARD 1983). Progeny SVP resemble viral cores but lack the prominent icosahedrally located projections observed in cores derived directly from virus particles. The mRNA transcribed from progeny SVPs is uncapped (ZARBL et al. 1980) and constitutes 80%–90% of the viral mRNA synthesized in infected cells. It seems that progeny SVPs with transcriptase activity may have a relatively short half-life and undergo further rearrangement to yield immature progeny viruses. This process appears to involve displacement of protein μ-NS and the acquisition of the outer capsid proteins and the 12 projections associated with the inner core. The last step in morphogenesis appears to be the addition of the major outer coat protein σ-3 (JOKLIK 1981) which inhibits the viral transcriptase activity (ASTELL et al. 1972).

Virus replication occurs in perinuclear inclusions which, late in infection, contain arrays of mature and immature progeny viruses embedded in a reticulum of "kinky" filaments comprised, at least in part, of vimentin (SHARPE et al. 1982).

Morphogenesis occurs within these inclusion bodies. Mature viruses accumulate in infected cells which eventually die and lyse. The bulk of progeny virions, however, remain associated with cell debris.

3 Bluetongue Virus Genes and Proteins

BTV consists of a core composed of 32 capsomers arranged in icosahedral symmetry, surrounded by an outer fibrillar coat (BOWNE and RITCHIE 1970; VERWOERD et al. 1972). The outer coat can be removed from the virion by treatment with mono- or divalent cations at appropriate pH and contains two proteins, VP2 and VP5 (HUISMANS et al. 1987a). The remaining core particle contains ten segments of virion RNA (VERWOERD et al. 1970), two major proteins VP3 and VP7, and three minor proteins VP1, VP4, and VP6. The capsomers consist perhaps entirely of VP7 (HUISMANS et al. 1987c; HYATT and EATON 1988).

Immunoelectron-microscopic studies showed that gold-labeled VP7 antibodies react weakly with fixed, intact BTV particles in the grid cell culture technique (HYATT et al. 1987; HYATT and EATON 1988). We have recently shown that unfixed BTV particles remain bound to the grid substrate during labeling experiments and bind large numbers of VP7 monoclonal antibodies. This indicates that VP7 is accessible on the outer surface of the virus (HYATT and EATON 1988). More information about the structural components of the virus is given in Chap. 2 of this volume.

In vitro translation of individual methylmercuric hydroxide-denatured double-stranded RNA segments of BTV has demonstrated the coding assignments of each RNA from BTV-1 (South Africa; MERTENS et al. 1984), BTV-17 (United States; GRUBMAN et al. 1983), BTV-10 (South Africa; VAN DIJK and HUISMANS 1988), and BTV-1 (Australia; EATON and HYATT 1989). The results show that in addition to the seven structural proteins, BTV codes for four additional proteins which have not been detected in virus preparations isolated from infected cells by freon extraction and purified by zonal centrifugation in sucrose and equilibrium centrifugation in cesium chloride (VERWOERD et al. 1972). Two of these nonstructural proteins, NS1 and NS2, are readily detected in BTV-infected cells. In contrast, NS3 and NS3a, which are closely related proteins generated by RNA 10, are difficult to detect in infected cells. Localization of NS1 and NS2 in virus-infected cells has provided some clues to the function of these two proteins (see Sect. 7). The role of NS3 and NS3a in virus replication is unknown. They are synthesized in small amounts late in infection and appear to be present in the soluble fraction of infected cells (VAN DIJK and HUISMANS 1988). Small amounts of NS1 and NS2 have been detected in virus particles purified using sodium lauryl sarcosine and dithiothreitol (MERTENS et al. 1987; EATON et al. 1988).

In contrast to NS2, which appeared to be present on the surface of the virus particle (MERTENS et al. 1987), NS1 was associated with core particles. Immunogold probing of core particles using NS1 monoclonal antibodies suggests that NS1 may project from the surface of the core particle (EATON et al. 1988). It is clear that neither NS1 nor NS2 are required for virus infectivity (VERWOERD et al. 1972). Their presence in the virus may be due to surface contamination of core or intact particles. Alternatively, the proteins may play a role in virus replication or morphogenesis (see Sect. 7).

4 Adsorption of Bluetongue Virus to Cells

Field isolates of BTV from infected ruminants or insects often replicate poorly in cultured cells. Adaptation by passage in embryonated chicken eggs and/or tissue culture selects viruses which replicate in primary cultures such as chicken embryo and lamb kidney and in a wide variety of cultured cells of insect and mammalian origin including BHK-21, Vero, MDBK, HeLa, SVP, and *Aedes albopictus* cells (HOWELL et al. 1967; McPHEE et al. 1982). The results outlined in this chapter were derived from studies on the interaction between BTV and susceptible cells in culture.

No information is available on the nature of the cell receptor to which BTV binds. However, analysis of the BTV-erythrocyte interaction has indicated that the virus binds to specific sialic acid-containing, serine-linked oligosaccharides in the glycophorins of human and a number of animal erythrocytes (EATON and CRAMERI 1989).

It is clear that virus binding to cells is mediated by the outer coat protein VP2. HUISMANS et al. (1983) showed that particles of BTV-10 lacking VP2, but containing the other outer coat protein VP5, were incapable of binding to BHK cells in suspension. Similarly, COWLEY and GORMAN (1987) used reassortants of BTV-20 (which agglutinated sheep erythrocytes only) and BTV-21 (which agglutinated sheep, bovine, human, and goose erythrocytes) to show that agglutinating ability correlated with the presence of VP2.

The virus adsorbs rapidly to susceptible cells at both 4° C and 37° C with maximum adsorption of BTV-10 to BHK cells and BTV-4 to mouse L cells occurring within approximately 20 min (HOWELL et al. 1967; HUISMANS et al. 1983).

5 Endocytosis

Investigation of many virus-cell systems by numerous researchers indicates that the most likely mechanism for virus entry into cells is direct internalization by plasma membrane penetration or utilization of the normal cellular process of receptor-mediated endocytosis followed by penetration of either the endosomal or lysosomal membrane. Viruses in two of the *Reoviridae* genera provide examples of each of these modes of uptake. Trypsin-activated, infectious rotaviruses enter cells by direct plasma membrane penetration (KALJOT et al. 1988). In contrast, reoviruses are efficiently transported into lysosomes, and it is there that the outer coat is partially hydrolyzed (STURZENBECKER et al. 1987) generating subviral particles which probably pass through the lysosomal membrane into the cytoplasm (DALES 1973).

There are several reports describing the presence of BTV in intracellular vacuoles shortly after infection (LECATSAS 1968; CROMACK et al. 1971). Since these studies were reported, much more information is available on the process of endocytosis and viral uptake via the endocytic pathway. Key features of this pathway are given below and micrographs of BTV at various stages of uptake and uncoating are seen in Fig. 1. The virus binds to a receptor on the plasma membrane (Fig. 1a, b) at a site which may ultimately be characterized by the presence of clathrin. The clathrin-coated membrane surface grows as a "coated pit" which eventually invaginates (Fig. 1c) and detaches from the cell surface, yielding a coated vesicle (Fig. 1d). This rapidly loses the clathrin coat (Fig. 1e). Vesicles may then fuse with (Fig. 1f), or fuse together to form, a larger electron-lucent, endocytic vesicle, the endosome (Fig. 1g). Endosomes lack the hydrolytic enzymes found in lysosomes but resemble the latter in having an acidic pH (HELENIUS et al. 1983). Most viruses appear to use the endosome as the site of entry into the cell cytoplasm. The low pH triggers a fusion activity in the glycoprotein spikes of enveloped viruses or a surface protein of nonenveloped viruses. This results in either fusion of the viral membrane with the limiting membrane of the vesicles or the direct penetration of the nonenveloped virus or its genome through the membrane. Most of the macromolecular contents of endosomes are ultimately transferred to secondary lysosomes which are hydrolase-rich, acidic, electron-dense vacuoles often located in the perinuclear region of the cell. Reovirus is unusual in its requirement for hydrolytic conditions to remove the virus outer coat and activate the virion transcriptase. Studies on the uptake of BTV into cells have to date revealed few virus-like particles in lysosomes (BROOKES unpublished). This suggests that BTV may enter the cytoplasm (Fig. 1h) by penetration of the endosomal membrane. The morphology of BTV particles within endosomes (Fig. 1f, g) indicates that the acidic conditions therein (approximately pH 5.0) have elicited either changes in the structure of the outer capsid layer or its removal from the virus. HUISMANS et al. (1987c) have shown that within 1 h of infection, BTV is converted to core particles (Fig. 1h) which appear to lack all of the VP2 and most of the VP5 of the outer capsid layer. It is interesting that VP2 can be removed from BTV in vitro at pH 5.0 (albeit in the presence of 0.2 *M* salt) (HUISMANS et al. 1987a). This finding is consistent with the morphological observations suggesting that BTV uncoating occurs in endosomes. The failure to remove VP5 in vitro at pH 5.0 raises the possibility that its removal may occur in the cell cytoplasm after release of a partially uncoated virus from the endosome. Removal of both outer capsid proteins is required to activate the virion transcriptase (VAN DIJK and HUISMANS

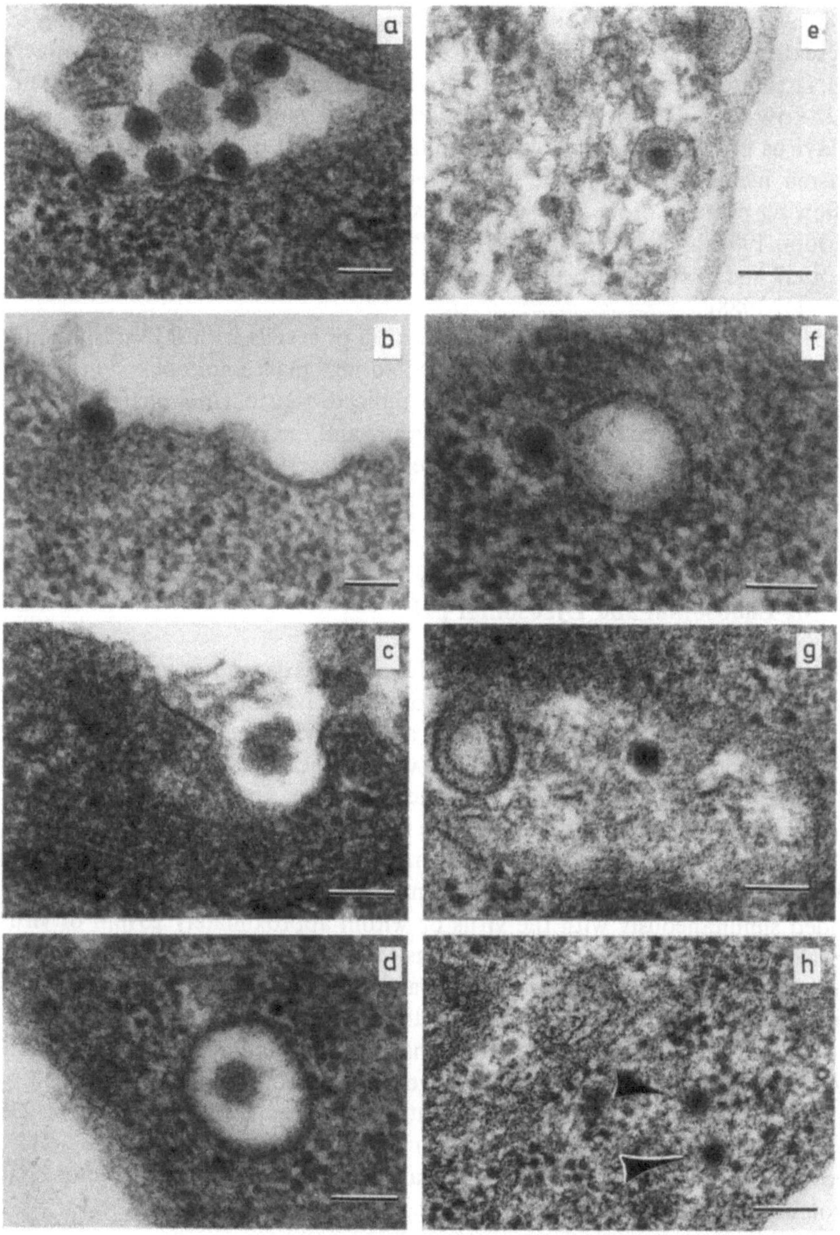

Fig. 1 a–h. Uptake of BTV by endocytosis and release of core-like particles into the cytoplasm. The *arrows* in H point to two core-like particles in the cytoplasm. *Bars*, 100 nm

1980). From these observations it would appear that a "fusion sequence" may be present in either partially or completely uncoated BTV particles which, at acid pH, facilitates their removal from the endosome and delivery to the cytoplasm. MACKOW et al. (1988) have recently identified a putative fusion sequence in rotavirus protein VP7 which may facilitate transport of the virus through the plasma membrane. However, similar sequences have not been found in the following proteins of the BTV-1 virus used in our morphological studies: VP3 (GOULD 1987a), VP2 (GOULD 1987b), VP5 (GOULD and PRITCHARD 1988), VP7 (GOULD and KATTENBELT, unpublished), NS1 (GOULD et al. 1988b), and NS3 (GOULD 1988).

HUISMANS (1970) has shown that infection of L cells with BTV results in a rapid inhibition of cell protein synthesis. No new macromolecular synthesis is required to induce this effect and data indicate that a virus coat protein is responsible. The speed with which this occurs and the lack of a requirement for replication suggest that a virus coat protein may exert a toxic effect at either the plasma or endosomal membrane.

5.1 Inhibition of BTV Replication by Modulators of Endosomal and Lysosomal pH

The release of virus from endosomes and lysosomes is dependent on the low pH in these organelles. Raising the intraendosomal and lysosomal pH by addition of lysomotropic weak bases such as NH_4Cl and methylamine or acidic ionophores, such as monensin and nigericin (OHKUMA and POOLE 1978), results in the failure of endocytosed virus particles to enter the cytoplasm. The effect of lysomotropic weak bases and acidic ionophores on the replication of BTV has been examined. The results indicate that NH_4Cl (20 m*M*), methylamine (20 m*M*), monensin (0.01 m*M*) or nigericin (0.005 m*M*) completely block virus replication when added simultaneously with the virus or within approximately 20 min of virus uptake (HYATT et al. 1989). This confirms that uptake of BTV by the endocytic route is required to generate a core particle with transcriptase activity and thereby initiate a productive virus infection.

HUISMANS et al. (1987c) have shown that intracellular core particles ultimately lose their VP7-containing capsomers to generate subcores which, at least in vitro, fail to transcribe RNA. This suggests that the transcriptase of each core particle may be active for only a limited period. In addition, the data suggest that, unlike reovirus, core particles do not reassociate with newly synthesized outer coat proteins and are not released from the cell as progeny virus.

6 Amplification of Infecting Virions

Following release from endosomes, core particles of BTV in the cytoplasm (Fig. 1h) become associated with a matrix that is similar in structure but less

large numbers later in infection (see Sect. 7.3). As seen in Fig. 2a, the matrix often appears to form at one side and in contact with the core particle but eventually spreads to surround it (Fig. 2b). These structures are undoubtedly VIB precursors and they are presumably formed by condensation of viral RNA transcribed from the core particle and viral proteins translated on ribosomes outside the VIB. Information on the protein content of VIB will be presented in Sect. 7.3. Particles with bishelled morphology are observed at the periphery of VIB starting at

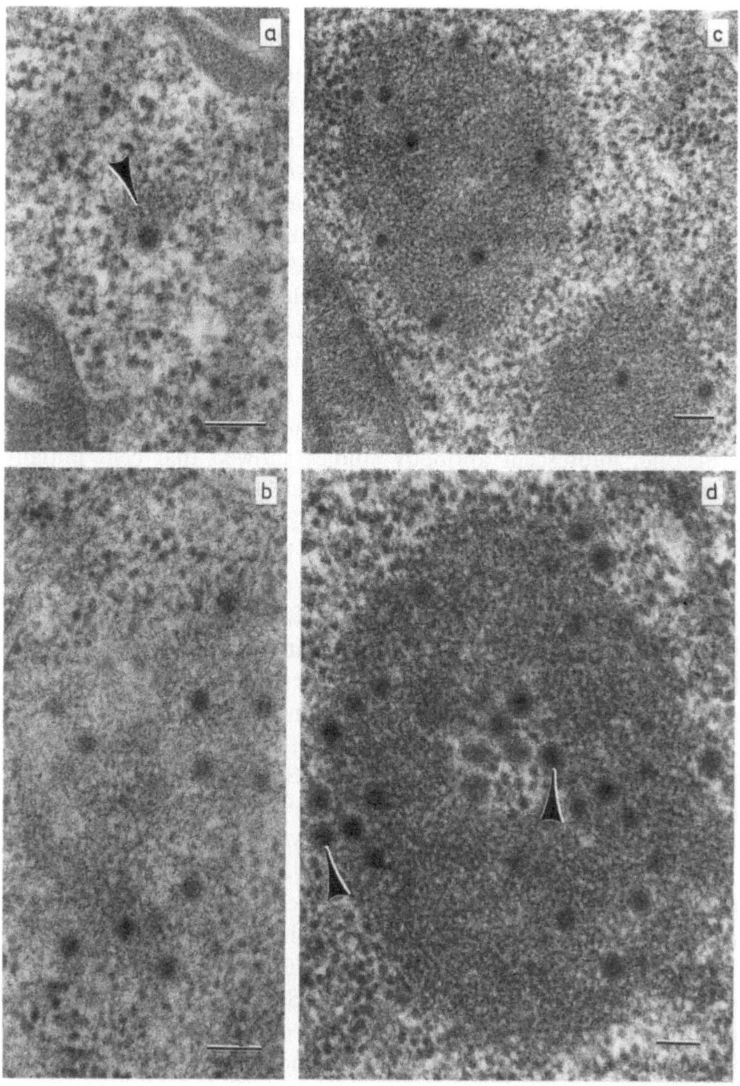

Fig. 2 a–d. Formation and development of virus inclusion bodies in BTV-infected cells. **a** and **b**, 4–6 h p.i.; **c**, 8–12 h p.i.; **d**, from approximately 14 h p.i. The first sign of VIB matrix formation is indicated by the *arrow* in **a**. *Arrows* in **d** point to bishelled virus particles. *Bars*, 100 nm

approximately 6 h postinfection (p.i.; Fig. 2b, c, d), indicating that virus particles are synthesized within these structures. Activation of the BTV transcriptase activity requires removal of both outer coat proteins VP2 and VP5 (VAN DIJK and HUISMANS 1980). The presence of an outer layer in progeny particles at the VIB periphery suggests that they contain at least some of the outer coat proteins and therefore may not be able to synthesize mRNA.

6.1 Superinfection of Virus-Infected Cells

Late in infection the cell contains numerous VIBs ranging is size from approximately 300 nm up to and exceeding 2 μm in diameter. The release of bishelled, probably transcriptase negative, virus particles from the small number of VIBs present early in infection raises the question of how the large increase in VIB number arises. There appear to be several possibilities. First, although the virus particles leaving VIBs early in infection have an outer layer, they may be transcriptionally active. They may also be converted in the cytoplasm into core particles which are capable of transcription and initiation of new VIBs. Alternatively, newly formed virus particles may be released from cells, reenter via the endocytic pathway, and reach the cytoplasm as transcriptionally active core particles. The latter possibility was suggested by two observations. First, electron-microscopic analysis of infected cells at 18–22 h postinfection (p.i.) revealed large numbers of virus particles in endocytic vacuoles (HYATT et al. 1989). Second, fluorescent antibody studies showed that cells initially infected at a multiplicity of infection of approximately 10 PFU/cell with a BTV variant (which failed to react with a specific VP2 antibody) supported the replication of wild-type BTV when it was added 6 h after the initial infection (HYATT et al. 1989).

To investigate the role of superinfection in the BTV replication cycle, virus was adsorbed to cells at 4° C for 1 h and at different times thereafter, at 37° C, 20 m*M* NH_4Cl was added. The titer of released virus, determined at 36 h p.i., showed that addition of NH_4Cl within the first 20 min essentially prevented virus replication. Addition at times up to 12 h p.i. decreased the titer approximately tenfold. A similar study with reovirus-infected cells (STURZENBECKER et al. 1987) indicated that addition of NH_4Cl after 30 min inhibited virus yield by only two to fourfold. Clearly NH_4Cl exerts the maximum effect when added very early in the replication cycle of both reovirus and BTV. Addition later has a smaller but nevertheless significant effect, especially in BTV-infected cells. The data support the idea that reinfection of BTV-infected cells (and perhaps reovirus-infected cells) by progeny virus contributes to the final virus titers. Nigericin exerts a similar effect in BTV-infected cells (HYATT et al. 1989).

To determine if the nigericin and NH_4Cl-mediated decrease of released virus titer was correlated with a fall in the intracellular concentration of virus antigen, cells on coverslips were infected with BTV and at 4 h p.i. nigericin or NH_4Cl was added. At 22 h and 46 h p.i. cells were probed by immunofluorescence using NS1-

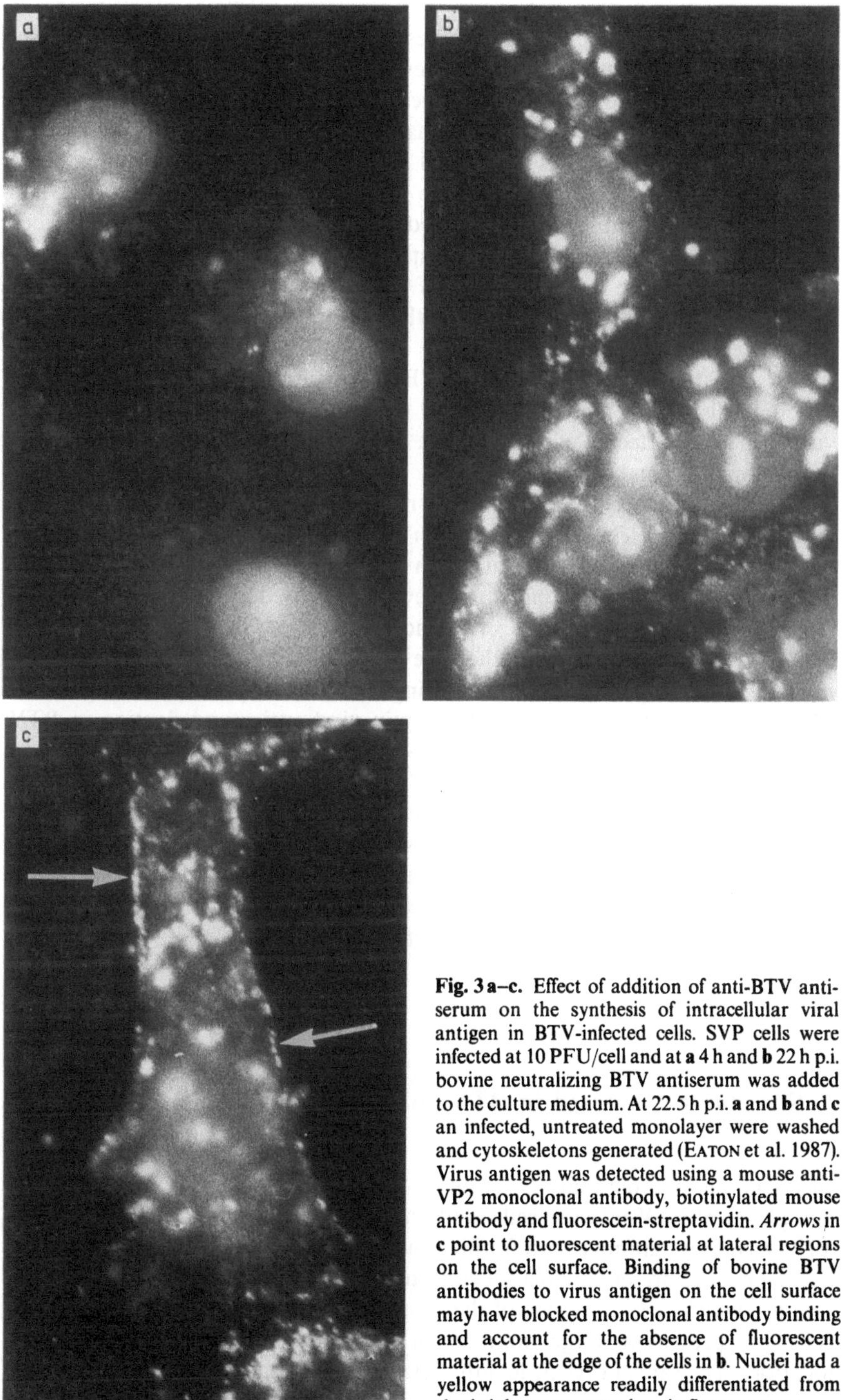

Fig. 3 a–c. Effect of addition of anti-BTV antiserum on the synthesis of intracellular viral antigen in BTV-infected cells. SVP cells were infected at 10 PFU/cell and at **a** 4 h and **b** 22 h p.i. bovine neutralizing BTV antiserum was added to the culture medium. At 22.5 h p.i. **a** and **b** and **c** an infected, untreated monolayer were washed and cytoskeletons generated (EATON et al. 1987). Virus antigen was detected using a mouse anti-VP2 monoclonal antibody, biotinylated mouse antibody and fluorescein-streptavidin. *Arrows* in **c** point to fluorescent material at lateral regions on the cell surface. Binding of bovine BTV antibodies to virus antigen on the cell surface may have blocked monoclonal antibody binding and account for the absence of fluorescent material at the edge of the cells in **b**. Nuclei had a yellow appearance readily differentiated from the bright green cytoplasmic fluorescence.

and VP2-specific monoclonal antibodies which bind to virus tubules and virus particles respectively (see Sect. 7). The results showed that at 22 h p.i. about 50% of the cells in treated cultures contained no detectable virus antigen and the remaining cells contained much less than that observed in untreated, infected cultures where all the cells fluoresced strongly. By 46 h p.i., the number of fluorescence-positive cells had risen to about 95% and the intracellular concentration of viral antigen had increased. At this time p.i. in untreated, infected cultures, no cells remained on the substrate (HYATT et al. 1989).

Further evidence for reinfection of cells by released progeny virions was sought by adding BTV antiserum to the culture medium of infected cells to neutralize released viruses and thereby prevent superinfection. The results in Fig. 3 show that addition of a bovine BTV-1-neutralizing antiserum to BTV-infected cells 4 h p.i. markedly reduced the amount of cytoskeleton-associated virus particles detectable at 22.5 h p.i. by immunofluorescence using a VP2 monoclonal antibody.

These results may be explained if it is assumed that a majority of virus particles leaving VIBs early in infection are incapable of transcription and are unable to initiate new VIB formation. In infected cells treated with NH_4Cl at 1 h p.i. for example, the development of new VIBs may depend on the infrequent release from established VIBs of unencapsidated particles which contain an active transcriptase. Increasing the time of infection under these circumstances may enhance the possibility of generating such transcriptase positive particles. This may account for the marked delay in viral antigen production in BTV-infected cells treated with nigericin or NH_4Cl at times after 1 h p.i.

In summary, in the absence of NH_4Cl, progeny virus made early in infection may be released from the cell (see Sect. 8) and reenter via endocytosis. The core particles liberated into the cytoplasm presumably initiate new VIB formation. An increase in the number of VIBs may raise the probability of synthesizing, within the cell, core-like particles with active transcriptases. The process whereby progeny virus particles reinfect cells effectively increases the multiplicity at which cells are infected with BTV.

7 Structures Observed in BTV-Infected Cells

Electron-microscopic analysis of thin sections has shown that late in infection the cytoplasm of BTV-infected cells contains three major viral specific structures. VIBs have both granular and fibrillar characteristics and are found throughout the cell but are present in greatest concentration in juxtanuclear locations. Virus tubules, which constitute one of the most unique features of orbivirus-infected cells, are present in large numbers in predominantly peri- or juxtanuclear locations. They are of varying lengths with a 68 nm diameter and appear to have a helical conformation with a 9 nm periodicity (HUISMANS and ELS 1979). Tubules isolated from the infected cells do not contain nucleic acid and the function of these structures is one of the most puzzling aspects of BTV replication. Unlike reovirus-infected cells, which contain large arrays of particles late in infection, electron-microscopic analysis of thin sections of BTV-infected cells rarely reveals similarly large aggregates of BTV particles. However, the use of monoclonal antibodies to VP2 in immunofluorescence experiments has shown that large VP2-containing aggregates are present in the cytoplasm and attached to the cytoskeleton (see below) of BTV-infected cells. Negative staining of intact, infected cells grown on electron-microscope grids has also revealed aggregates, which may be quite large, underlying the infected cell membrane. The involvement of such aggregates in virus morphogenesis and release is discussed in Sect. 8.

7.1 Attachment to Intermediate Filaments

Eucaryotic cells contain an extensive filamentous network referred to as the cytoskeleton and which is revealed by treatment of cells with NP40 or Triton-X100. Such nonionic detergents solubilize the cell membrane and release soluble cytoplasmic components. The three major classes of protein filament in the cytoskeleton are the actin-based microfilaments (approximately 5 nm diameter), the tubulin-containing microtubules (approximately 25 nm diameter), and the intermediate filaments (10 nm diameter) whose composition depends on the cell type (EATON and HYATT 1989).

Early recognition of a possible association between double-stranded RNA viruses and the cytoskeleton resulted from the work of DALES (1973) who analysed thin sections of so-called virus factories in reovirus-infected cells. This revealed that virus particles were associated with microtubules and enmeshed in masses of dense, twisted, or kinky filaments. Subsequent studies by BABISS et al. (1979) confirmed that reovirus type 1 had a high affinity for microtubules even in vitro. More recently SHARPE et al. (1982) have shown that there is a specific disruption of the intermediate filament network in reovirus-infected cells and a relocalization of the intermediate filament protein vimentin in reovirus inclusion bodies.

Preliminary evidence for a cytoskeletal role in BTV replication was obtained when it was shown that BTV antigens which react with both polyclonal and BTV monoclonal antibodies remain associated with the cytoskeleton following treatment of infected cells with NP40 (EATON et al. 1987; EATON and HYATT 1989). In addition, analysis of the detergent-soluble cytoplasmic fraction and the insoluble nuclear-cytoskeletal fraction of BTV-infected cells labeled with [^{35}S]-methionine revealed that a high proportion of the individual structural and nonstructural proteins remain in the cytoskeletal fraction.

Visual evidence for the presence of virus-specific structures on the cytoskeleton was obtained using whole mount electron microscopy of the cytoskeleton. Cells grown on grids were infected with BTV and at 20 h p.i.

cytoskeletons were prepared and fixed by simultaneous treatment with NP40 and 0.1% glutaraldehyde (HYATT et al. 1987). Cytoskeletons were postfixed in osmium tetroxide, dehydrated in acetone, and critical-point dried from carbon dioxide. The detergent-insoluble skeletal networks of BTV-infected cells were characterized by the presence of dense areas in juxtanuclear positions. Within these areas and in other perinuclear locations were large numbers of VIBs and virus tubules. The VIBs appeared to be approximately spherical structures. Virus tubules appeared in groups and were often found lying parallel to neighboring filaments. Individual tubules appeared hollow, approximately 30 ± 2 nm in diameter and ranged in length from 300 to 400 nm. Virus-like particles were identified on the basis of size (55 nm) and their approximately hexagonal shape, and were found either singly, in rows, or in aggregates. BTV-like particles were also observed outside on the cells surface or leaving the cell (see Sect. 8).

The viral-specific structures bound to filaments with a diameter of approximately 10 nm. This effectively ruled out the larger (25 nm) microtubules as the site of binding. To determine if virus-like particles, VIBs or tubules bound to intermediate or microfilaments, three monoclonal antibodies specific for each of these structures (see Sects. 7.2, 7.3, and 7.4) were used to probe cytoskeletons of BTV-infected cells following treatment with the microfilament-disrupting drug cytochalasin B. Although the drug altered the distribution of fluorescence with each of the three monoclonal antibodies, it did not eliminate it from cytoskeletons. Thus, the binding of viral-specific structures was not dependent on intact microfilaments. Similar results were obtained with the microtubule-disrupting drug colchicine (EATON et al. 1987).

These results implicated intermediate filaments as the site of binding. Direct evidence for this came from two sources. First, virus-like particles, VIBs and tubules were associated with the intermediate filament array which formed around the nucleus in infected cells treated for prolonged periods with colchicine (EATON et al. 1987; EATON and HYATT 1989). In particular, virus-like particles were associated in long trains with the intermediate filaments (Fig. 4). Second, virus particles were found to be associated with filaments which reacted with colloidal gold-labeled anti-vimentin antibodies (EATON et al. 1987; EATON and HYATT 1989).

In summary, these data show that at least a proportion of VIBs, tubules, and virus-like particles in BTV-infected cells is cytoskeleton-associated. VIB and tubule-specific monoclonal antibodies have been used to determine if all of these structures are bound to the cytoskeleton or if some are found in the soluble cytoplasmic fraction of infected cells. Comparison of the patterns and intensity of fluorescence generated by methanol-fixed intact cells with those of cytoskeletons following probing with a VIB-specific antibody suggests that the vast majority, if not all, of the VIBs are cytoskeleton-associated. In contrast, results with a tubule-specific monoclonal antibody suggest that some tubules may not be associated with the cytoskeleton. Electron-microscopic examination confirms the presence of tubules in NP40-generated cytoplasmic extracts of infected cells (HYATT and EATON 1988).

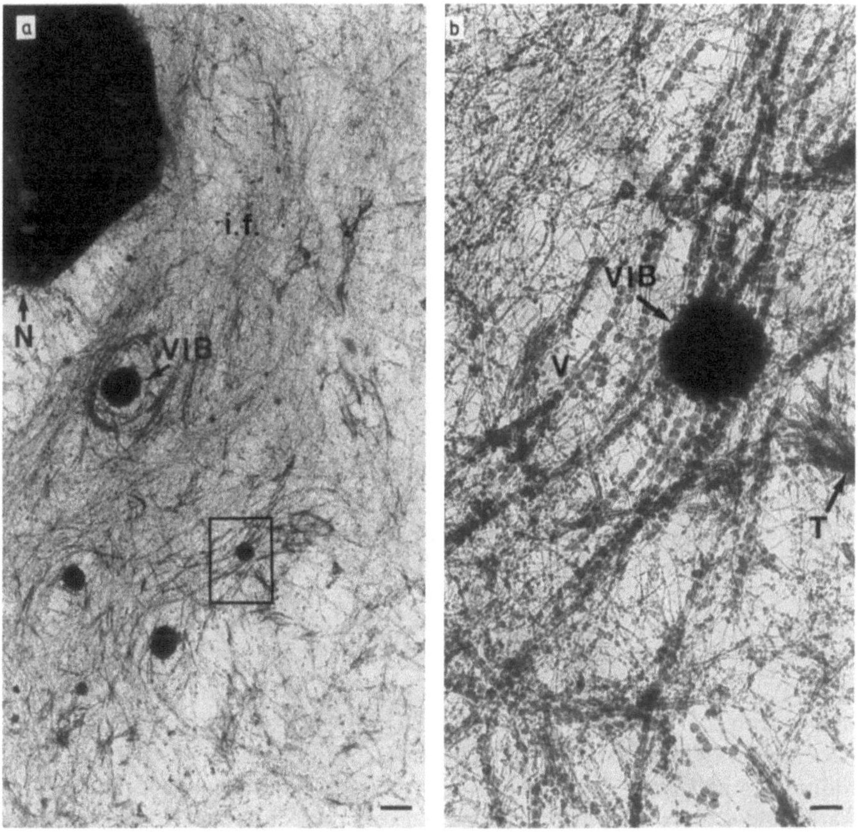

Fig. 4 a, b. Effect of colchicine on the cytoskeletal network of BTV-infected cells. Colchicine was added immediately after virus infection and cytoskeletons prepared at 22 h p.i. *N*, nucleus; *T*, tubule; *V*, virus particle; *VIB*, virus inclusion body; *i.f.*, intermediate filament. *Rectangle* in **a** is enlarged in **b**. *Bars*, 1 μm in **a** and 200 nm in **b**.

7.2 Virus Tubules

HUISMANS and ELS (1979) have purified tubules from BTV-infected cells and shown them to consist predominantly, if not entirely, of the nonstructural protein NS1. Probing of cytoskeletons of BTV-infected cells using monoclonal antibodies with known specificity for virus proteins confirms that tubules contain NS1 and reveal the presence of VP3 and VP7 (EATON et al. 1988; HYATT and EATON 1988). The fact that tubules isolated from infected cells and purified by sucrose density gradient centrifugation contain only NS1 in detectable amounts, suggests either that only small quantities of structural proteins are present or that during purification the structural proteins are removed.

7.3 Virus Inclusion Bodies

The presence of dense, core-like structures within VIBs and virus-like particles at the periphery suggest that it is within these bodies that virus morphogenesis occurs. The identification of RNA in VIBs as detected with acridine orange stain (Bowne and Jochim 1967) is consistent with this belief. Examination of thin sections of VIBs indicates that most of the putative BTV precursor particles within the matrix lack a visible outer layer (Fig. 1c). In contrast, at the VIB periphery, the majority of virus-like particles are bishelled. Previous reports have described, and we have observed, apparently complete, bishelled virus particles within VIBs (Fig. 1d). It must be recognized, however, that VIBs are not necessarily spherical or elliptical in shape, and that they may contain large invaginations. Virus particles at the periphery of VIBs but within such "tunnels" may appear in thin sections to be within the VIBs. The fact that apparently complete bishelled virus particles within VIBs are often present in areas of decreased density supports this contention (Fig. 1d).

Immunofluorescent studies showed that cytoskeleton-associated VIBs (identified as approximately circular, dense structures under phase contrast) bound antibodies to both VP3 and VP7. Indication that these proteins were present not only at the periphery but also within the VIB matrix came from studies in which Lowicryl sections of whole cells were probed with anti-VP3 and anti-VP7 monoclonal antibodies (Hyatt and Eaton 1988). Anti-VP7 antibodies labeled with colloidal gold reacted with the VIB matrix but did not appear to be associated with the dense virus-like structures in VIBs. This suggests that VP7, in addition to its known presence within virus cores, constitutes part of the VIB matrix and is there in high concentration. Lowicryl sections of VIBs labeled very weakly (one or two gold particles per VIB) with anti-VP3 antibody. It seems unlikely that this protein is located solely on the periphery of VIBs, as indicated by immunofluorescence experiments. Thus, the low but specific level of anti-VP3 labeling indicates the presence of VP3 within VIBs. In contrast, gold-labeled antibodies to VP2 fail to label the internal VIB matrix. In addition, the immunofluorescent probing of cytoskeletons from infected cells shows that VIBs label very weakly with anti-VP2 monoclonal antibodies. The labeling, if it is present, is limited to one or two pin-point fluorescent foci (Gould et al. 1988a).

The nature of the interaction of VP2 with VIBs was investigated by probing cytoskeletons with an anti-VP2 antibody conjugated directly to gold. Following embedment, thin sections were cut and examined in the electron microscope. The results show that anti-VP2 reacted only in locations where virus-like particles appeared to be leaving the VIBs (Gould et al. 1988a; Hyatt and Eaton 1987). Thus, VP2 may be added to developing virus particles, perhaps cores or core-like particles, at the periphery of the VIBs. However, as pointed out below, virus particles outside the cell and cytoskeleton-associated virus-like particles distal to VIBs, are labeled with anti-VP2 antibody more intensely than the particles apparently maturing at the VIB periphery. The difference is not due to an inability of the gold-labeled antibody to diffuse efficiently into the cytoskeleton and reach the VIB (HYATT 1988). There are two explanations which account for these observations. First, VP2 may be present in VIBs in a form undetectable by the anti-VP2 antibodies used to date. The increase in the reactivity of cytoskeleton-associated virus-like particles with gold-labeled anti-VP2 antibody may reflect a reorganization in the outer coat of the virus, after release of the virus from VIBs, with a subsequent increase in accessibility of the appropriate epitope on VP2. Alternatively, although some VP2 is added to core-like particles at the VIB periphery, the remainder may be added after the virus-like particles leave the VIBs, either in the cytosol or following binding to the cytoskeleton. Analysis of cells infected with a neutralization-resistant variant of BTV-1, V35B2, provides evidence in support of the suggestion that VP2 is added to core-like particles at the periphery of and outside VIBs (GOULD et al. 1988a). In cells infected with this variant, VIBs appear surrounded by particles whose size and morphology are consistent with their identification as core particles. Not surprisingly, these particles label very weakly with anti-VP2 antibodies, whereas the virus-like particles elsewhere on the cytoskeleton and those released from the cell label strongly with the anti-VP2 monoclonal antibody.

Monoclonal antibodies to NS2 react with the surface of VIBs in immunofluorescence and immunogold labeling procedures. Labeling of the internal matrix of VIB in Lowicryl sections of whole cells was also achieved with anti-NS2 monoclonal antibodies. NS2 is phosphorylated and has been shown to bind to single-stranded RNA (HUISMANS et al. 1987b). Its presence within VIBs is consistent with a role for this protein in RNA synthesis or sequestration of viral single-stranded RNA molecules prior to virus morphogenesis. Unlike monoclonal antibodies to NS2, NS1 antibodies fail to react with the matrix of VIB in Lowicryl sections. Some but not all NS1 antibodies do, however, react in immunofluorescence tests, i.e., with the surface of the VIBs. To delineate the nature of NS1 in this location, cytoskeletons of infected cells were incubated sequentially with NS1 antibody 20E6/A4 and protein A-gold. Following embedment and thin sectioning, VIBs were found to be nonuniformly labeled with gold. The label appeared to be localized around and in the immediate vicinity of virus particles leaving VIBs. Importantly, gold was observed in association with fibrillar arrays around virus particles observed at the edge of the VIBs (EATON et al. 1988). There are two possible sources for the NS1-containing material at the VIB periphery. First, NS1 may be present in VIBs in an immunologically undetectable form and accompany the virus core particles which mature from there. Second, NS1 may be added to the developing virus particles at the VIB periphery. This would account for the failure to detect NS1 inside VIBs and its presence at the VIB periphery in association with virus particles.

We have previously shown that some gold-labeled anti-NS1 monoclonal antibodies react with purified BTV core particles (EATON et al. 1987). This is evidence for an association which may begin at the VIB periphery. The localization of VP2 and NS1 monoclonal antibodies at the periphery of VIBs, where virus particles appear to be leaving, is indirect evidence for the role of this nonstructural protein in morphogenesis.

The presence on the cytoskeleton of virus particles which lack the fibrillar material raises the possibility that following release of the virus-fibril complex from the VIB, a late step in morphogenesis occurs, after which the fibrillar material may be removed and the virus released from the cytoskeleton into the cytosol. In the event that this is correct, what happens to the fibrillar material? It appears unlikely that the material is broken down into constituent soluble proteins because little, if any, soluble NS1 is detected in the cytoplasmic fraction of infected cells. An attractive possibility is that the NS1-rich fibrillar material condenses to form tubules. In such a model, tubules would be the repository of NS1 that has been utilized in a prior stage of virus morphogenesis.

Electron-microscopic examination of BTV-infected cells indicates that large numbers of tubules are present throughout the cell as early as 6 h p.i. and before the appearance of significant numbers of VIBs and progeny virus particles (BROOKES 1989). This observation suggests that a proportion of the NS1 made in infected cells may not be involved in morphogenesis and may also undergo condensation to form tubules. Perhaps NS1 is able to fulfill its function in morphogenesis only during or shortly after its synthesis on polysomes.

7.4 Virus and Virus-Like Particles

Virus and virus-like particles in BTV-infected cells are identified in thin section on the basis of their diameter (approximately 65 nm) and structure which encompasses a dense nucleoid (approximately 40 nm diameter) surrounded by a ring of electron-translucent material. In whole mount electron microscopy, virus-like particles have a characteristic hexagonal appearance and a diameter of approximately 55 nm. Particles with these criteria and the ability to react with monoclonal antibodies to VP2 are observed in several locations in BTV-infected cells. In addition to those released from the cell and adsorbed to the electron microscope grid in the grid cell culture technique, virus-like particles are associated with the intracellular cytoskeleton, the cortical layer of the cytoplasmic matrix underlying the cell membrane, and the surface of the infected cell (HYATT et al. 1989). Glutaraldehyde-fixed, intact cells and cytoskeletons have been probed with anti-VP2 antibody and the pattern and intensity of fluorescence compared. The results indicate that a majority of the viruses and virus-like particles associated with the cytoskeleton of infected cells are in intracellular locations.

Shearing of the nuclear-cytoskeleton fraction of BTV-infected cells in a Dounce homogenizer releases high titers of infectious virus. Infectious virus particles are also present in the cytosol of infected cells and are released by treatment of cells with NP40. Virus particles released during infection into the culture medium and those in both the cytosol and cytoskeletal fractions have been partially purified by sedimentation through 40% sucrose columns. Comparison of the virus titer and the number of particles detected by electron microscopy indicate a particle to infectivity ratio of 5–10:1 for each virus preparation. This finding and the fact that a majority of the cytoskeleton-associated virus particles are intracellular suggests that a significant proportion of the virus-like particles associated with the intermediate filament network of the cytoskeleton are infectious.

Electron-microscopic examination has revealed differences in the size of partially purified released, cytosol, and cytoskeletal viruses from cells at 22 h p.i. (HYATT et al. 1989). Cytoskeleton-associated virus particles are heterogeneous and display a range of diameters from approximately 64 to 74 nm. In contrast, released and cytosol virus particles appear more uniform with a majority of particles having a diameter of 64 to 67 nm.

To further investigate the properties of cytoskeletal-associated virus particles, viruses in the cytosol, and released viruses, we have quantitated their ability to react with a gold-labeled VP2 monoclonal antibody. To examine released viruses, cells were infected on electron-microscope grids and at 22 h p.i. virus particles adsorbed to the grid outside the cells were fixed with 0.1% glutaraldehyde prior to addition of gold-labeled antibody. Binding of antibody to cytoskeletal-associated virus particles was determined after treatment of infected cells with NP40 and glutaraldehyde. To probe virus particles in the cytosol of infected cells under conditions identical to those used for the immunological analysis of cytoskeleton-associated and released virus particles, use was made of the observation that following lysis of infected cells with NP40 a proportion of the virus particles in the cytosol bind to the grid substrate in close proximity to the cell. The previously released virus particles can be differentiated from the cytosolic virus by prior labeling of the former (and the intact unfixed cell) with a VP2 monoclonal antibody conjugated with a 6 nm gold probe. Following cell lysis with NP40, the newly released particles from the cytosol which have adsorbed to the grid and are unlabeled are probed with the same monoclonal antibody conjugated to a larger gold probe. The results indicated that there was a heterogeneous distribution of gold probes on the three virus populations, especially those derived from the cytosol. More importantly, the binding profiles suggested that the populations differed in their capacity to bind antibody. Surprisingly, viruses from the cytosol bound most and released viruses bound least gold-labeled anti-VP2 antibody (HYATT et al. 1989). There are several possible explanations which account for differences in monoclonal antibody-binding capacity of the three virus populations. First, cytoskeleton-associated and cytosolic virus particles contain more VP2 than released virus and this extra VP2 is removed during morphogenesis. Alternatively, there is no difference in the amount of VP2 present, but virus particles undergo a conformational rearrangement which obscures the reactive epitope on some VP2 molecules.

It is not clear if antibodies are able to neutralize both released viruses and those in the cytosol with equal efficiency. Failure to effectively neutralize virus particles liberated as a result of cell death and lysis may account for the reported instances of coexistence of infectious virus and neutralizing antibody in the serum of animals infected with BTV (see Chap. 5, this volume).

8 Release of Virus from Infected Cells

The mechanism whereby BTV is released from infected cells has been the subject of some speculation. Reoviruses tend to accumulate in cytoplasmic inclusion body-associated arrays and are not released until cell lysis late in infection. Similar large-scale paracrystalline arrays of virus particles have not been observed in BTV-infected cells. BTV may be released from cells in a number of ways. First, virus may appear in the culture medium following the death and subsequent lysis of infected cells. However, the release of virus within 4 to 5 h of infection of MDBK cells with BTV serotype 8 (BTV-8) and the thousand-fold excess of released over cell-associated virus up to 15 h p.i. in this system (Cromack et al. 1971) suggests that other mechanisms may be operative. In other BTV-cell systems a high proportion of virus remains cell associated (Howell and Verwoerd 1971). Although orbiviruses are observed predominantly as naked virus particles, there are many reports which describe the presence of virus particles with lipoprotein envelopes (Owen and Munz 1966; Els and Verwoerd 1969; Bowne and Ritchie 1970; Foster and Alders 1979). These are presumably due to virus "budding" from infected cells and viruses undergoing this mode of release have been observed (Bowne and Jochim 1967; Bowne and Ritchie 1970). Similar observations have been made for many other orbiviruses (Verwoerd et al. 1979). A third mode of virus release was suggested by a report which describes released virus particles situated close to "discontinuities" in the plasma membrane (Lecatsas 1968).

Using a variety of techniques including immunofluorescent and immunoelectron microscopy and the grid cell culture technique we have investigated the release of BTV-1 from infected cells. The presence of VP2 on the surface of BTV-infected cells 18 to 24 h p.i. has been revealed following binding of gold-labeled anti-VP2 antibody to glutaraldehyde-fixed, intact cells and enhancement of the gold with silver (Fig. 5).

Most, if not all, of this viral protein is linked to the cytoskeleton because immunofluorescent experiments have shown that anti-VP2 antibody added to unfixed cells at 37°C for 30 min prior to cell lysis remains associated with the cytoskeleton following addition of nonionic detergent. Electron-microscopic observations of thin sections and cytoskeletons of infected cells labeled with gold-tagged anti-VP2 antibody indicate that the VP2 on the cell surface is present as part of virus particles.

Negative staining of intact, infected cells labeled with anti VP2 antibody reveals virus particles both under the cell membrane (and therefore inaccessible

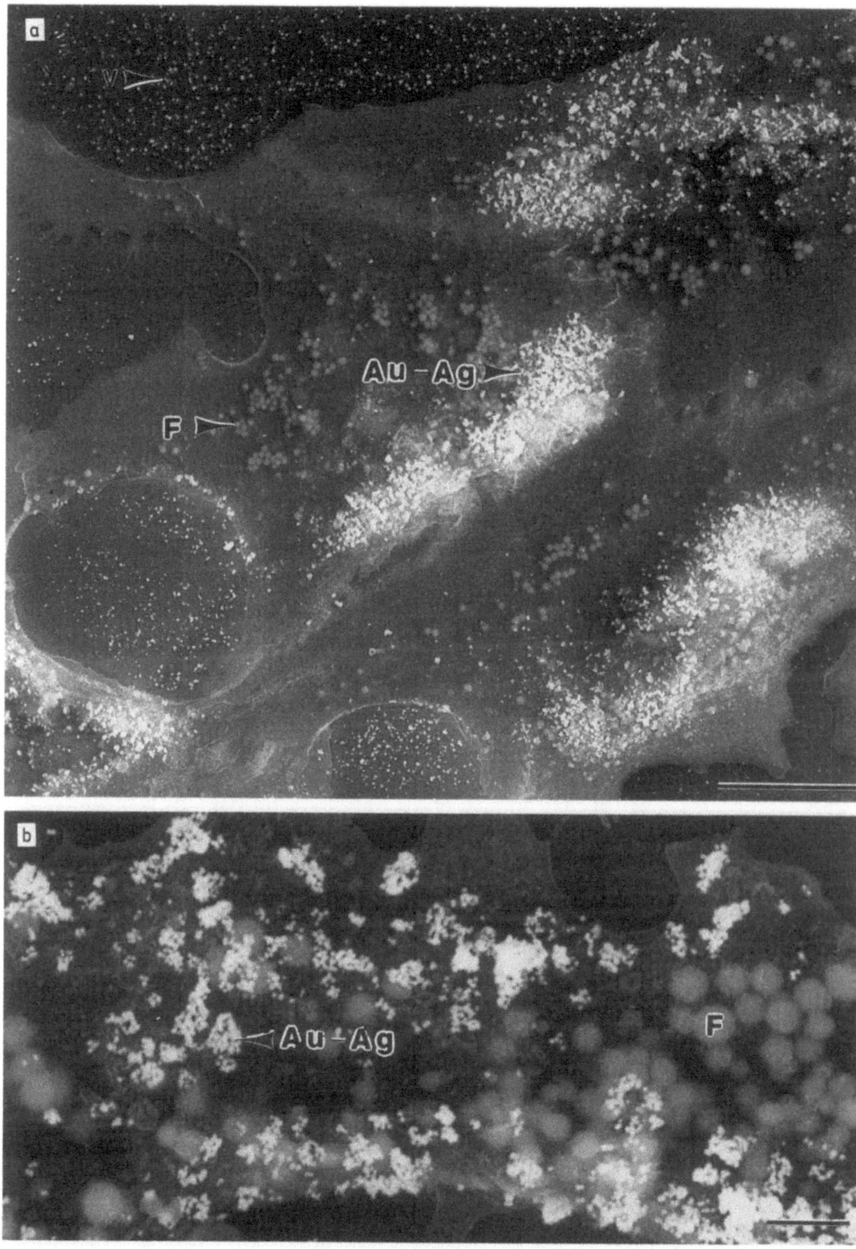

Fig. 5a,b. Immunoscanning electron micrograph of BTV-infected cells labeled with VP2 monoclonal antibody complexed with 12-nm gold particles and silver enhanced. *V*, virus particles bound to the substrate outside the cell (HYATT et al. 1989); *F*, fat droplet. *Bars* in **a** and **b** represent 10 μm and 1 μm respectively

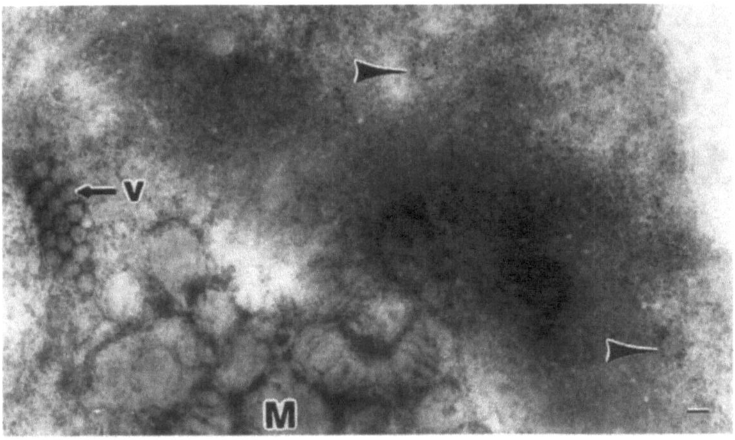

Fig. 6. Negative contrast immunoelectron micrograph of BTV-infected cells. Prior to negative staining infected cells grown on electron microscope grids were labeled with VP2 monoclonal antibody conjugated to 12-nm gold particles. *V*, virus particles underlying the membrane and unlabeled; *M*, mitochondria. *Arrowheads* point to labeled virus particles on the cell surface

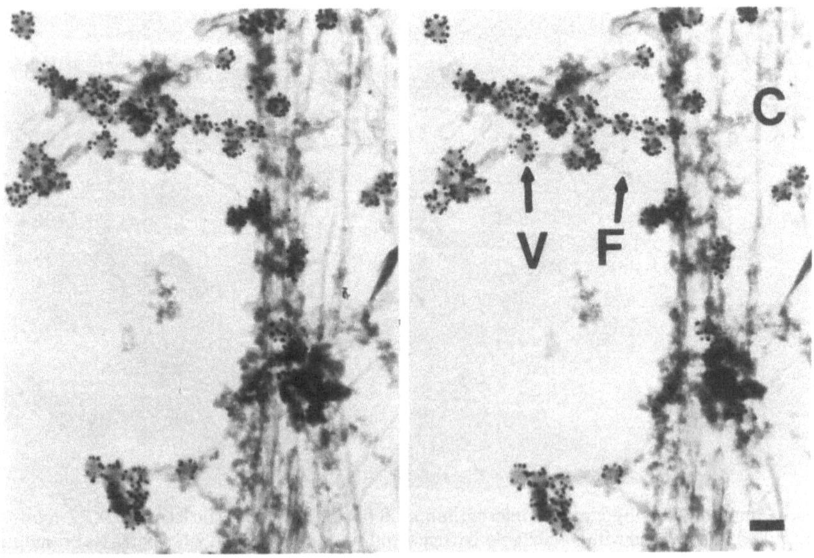

Fig. 7. Release of virus particles from infected cells. Cytoskeletons of BTV-infected cells were probed with gold-labeled VP2 monoclonal antibody and carbon coated. Stereomicrograph showing virus particles outside the cell but attached to the cortical layer underlying the cell membrane. *V*, virus particle; *F*, filament; *C*, cytoskeleton. *Bar*, 100 nm

to gold-labeled antibody) and on the cell surface (Fig. 6). Virus aggregates and single virus particles on the cell surface retain an association with the cortical layer of the cytoskeleton following cell lysis (Fig. 7).

Large numbers of virus particles were observed to bud through the cell membrane (Fig. 8a). However, most virus particles attached to the grid substrate, outside the cell, were not enveloped. This may be due to the failure of large numbers of enveloped viruses to adsorb to the grid substrate. Enveloped viruses on the other hand may be unstable and lose the membrane soon after release from the infected cells. Put another way, BTV particles may "escape from" or penetrate

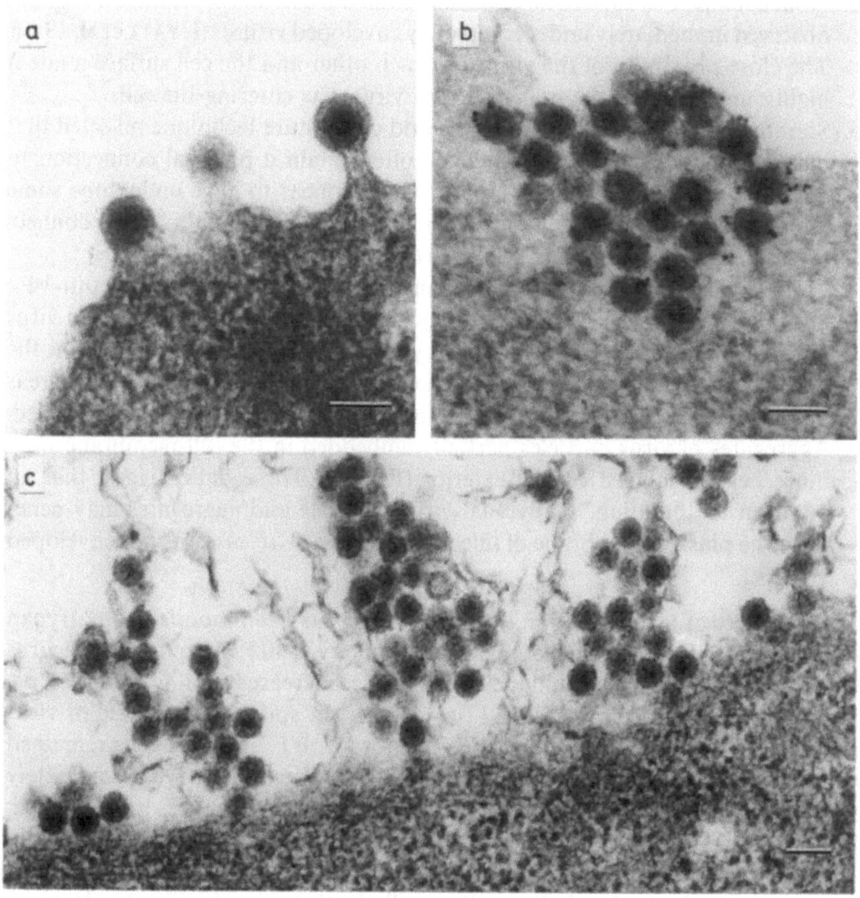

Fig. 8 a–c. Virus release from infected cells. Thin sections of BTV-infected cells showing **a** virus "budding", **b** an aggregate of nonenveloped virus particles penetrating the cell membrane and **c** membrane-like material and nonenveloped virus particles at the cell surface. Virus particles in **b** were labeled during a 1-h incubation of unfixed cells at 37° C with a VP2 monoclonal antibody conjugated to 6 nm gold. *Bars*, 100 nm

their surrounding lipid envelope. Some evidence in favor of the latter suggestion comes from the finding of large accumulations of membrane-like material outside infected cells (Fig. 8c).

However, it is clear that not all viruses leave the infected cell by a process of "budding." A large number of virus particles which appear to be partially embedded in the plasma membrane are observed without an overlying envelope (Fig. 8b). There are several observations which lead us to conclude that these particles are leaving rather than entering the cell:

1. Following addition of BTV to previously uninfected cells, virus uptake was observed to occur only by endocytosis and not by direct membrane penetration (Brookes 1989).
2. Nonenveloped virus particles, partially embedded in the membrane, have been observed immediately under completely enveloped viruses (Hyatt et al. 1989). The close proximity of the viruses to each other and the cell surface made it highly unlikely that the nonenveloped virus was entering the cell.
3. Staining of virus-infected cells in the grid cell culture technique revealed that naked virus particles outside the cell often retain a physical connection to localized areas of the cell membrane which appear to have undergone some alteration, thus permitting an increased penetration of stain. These connections were also observed in cytoskeletons of infected cells (Fig. 7).
4. Following incubation of living, virus-infected cells with gold-labeled anti-VP2 antibody for 1 h at 37° C, the vast majority of intracellular gold-labeled virus particles were observed within endocytic vesicles. Virus particles within the cell and those underlying the cell membrane remained unlabeled, whereas nonenveloped viruses outside, on the cell surface, were heavily labeled. Aggregates of virus particles partially embedded in the cell membrane were more heavily labeled on the exterior (Fig. 8b). These data suggest that, in addition to "budding," individual virus particles and aggregates may penetrate the plasma membrane of infected cells and be released as nonenveloped viruses.

The effect of virus infection on cell viability has been monitored by trypan blue exclusion. Following infection at a multiplicity of 10 PFU/cell, released virus was first detected at 6 h p.i. and the rate of release increased rapidly up to 12 h p.i. Maximum release occurred from 12 to 24 h p.i. In spite of the release of virus particles by penetration of the plasma membrane, BTV-infected cells remained capable of excluding trypan blue. As estimated by dye exclusion, approximately 95% of cells remained viable in both control and infected cultures up to 24 h p.i. After this time, infected cells started to detach, but even at 30 h p.i. approximately 75% still remained on the substrate. Approximately 80% of these remained viable. These observations also indicate that cell death and subsequent release of infectious virus do not contribute significantly to the titer of released virus at least up to 24 to 30 h p.i. The integrity of the membrane is also manifest by the ability of infected cells to retain ^{51}Cr. We have observed no difference in the rate of release of radioactivity from BTV-infected or control cells which had been labeled for 1 h with ^{51}Cr prior to virus infection. These cultures were observed for a period of 24 h after infection (HYATT et al. 1989).

Penetration of the plasma membrane by infecting icosahedral viruses has been observed with rotaviruses (KALJOT et al. 1988), infectious subviral particles of reovirus (BORSA et al. 1979), and adenoviruses (BROWN and BURLINGHAM 1973). Binding of adenoviruses and picornaviruses to the plasma membrane results in a destabilization or rearrangement of the virus particles. Adenovirus becomes DNase-sensitive due to the loss of some penton base and fiber proteins (OGIER et al. 1977). Poliovirus loses protein VP4 (FENWICK and COOPER 1962). Interaction of poliovirus with isolated plasma membrane preparations results in the loss of VP4 and VP2 (DE SENA and TORIAN 1980). Thus, the destabilization of virus particles at the plasma membrane outside the cell may be the first step in a virus-uncoating process that ends in the endosome or lysosome with the exposure of hydrophobic regions of the particle which facilitate entrance into the cytoplasm. Alternatively, rearrangement of the virus particles at the plasma membrane may precede or occur concomitantly with virus penetration through the membrane (BROWN and BURLINGHAM 1973). It is tempting to speculate that egress of BTV particles from infected cells by membrane penetration may occur in association with a reorganization of the outer capsid layer of the virus. The results presented in Sect. 7.4 indicate that BTV particles released from infected cells differ from intracellular particles in their ability to bind gold-labeled VP2 antibody. This difference may reflect either loss of VP2 during release or a rearrangement of the outer coat which obscures a proportion of the reactive epitopes on VP2. It must be noted that the size of BTV particles probably precludes unassisted movement through the actin-rich cortical layer underlying the cell membrane (HARTWIG et al. 1985). Presumably, there may also be a mechanism to facilitate this process. Whether this has any bearing on the structure of the neighboring plasma membrane is not clear.

9 Conclusions

In this chapter the replication of BTV has been discussed from a predominantly morphological perspective. Several facets of the virus replication cycle such as adsorption, uncoating, late stages in morphogenesis, and virus release from the cell are particularly amenable to investigation using electron microscopy. In addition, the use of a panel of monoclonal antibodies to virus structural and nonstructural proteins employing a variety of immunofluorescent and immunoelectron-microscopic procedures has provided new information on the cytoskeletal location and protein content of virus-specific structures synthesized in infected cells.

The data have permitted some conclusions to be drawn and encouraged us to speculate further on the mechanism of addition of outer coat proteins to BTV particles during morphogenesis, the effect of lysomotropic agents on virus infection, and the processes whereby virus particles are released from infected cells. These topics are dicussed within the following summary of BTV replication.

Following adsorption of cells, BTV is taken up by endocytosis. The outer coat protein VP2 is removed in endosomes. VP5 is probably removed there also prior to release of the core-like particles into the cytoplasm. In contrast to BTV, the outer coat of reovirus is only partially uncoated in the lysosomes. In the cytoplasm core-like particles of BTV bind to intermediate filaments and transcribe virion RNA. Translation of viral RNA generates proteins which appear to condense probably with single-stranded RNA to form a matrix which ultimately surrounds the core particle, thereby generating a VIB. It is within the VIB matrix that new core-like particles are made. The bishelled structure of particles at the VIB periphery suggests that at least a proportion of the outer coat proteins is added in that location. Some VP2 appears to be added there and the remainder may follow when the virus leaves the VIB. The site of VP5 addition to particles is unknown and the details of core particle morphogenesis within the VIBs are obscure. The size and density of the inner capsid observed in thin sections and its resemblance to that of virus cores suggests that the particles leaving the VIBs contain double-stranded RNA. Therefore, BTV replicase particles, i.e., those containing single-stranded RNA, may be found within the VIBs. Unlike reovirus-infected cells, where complete and incomplete particles coexist in inclusion bodies, the data indicate that late steps in BTV morphogenesis occur outside VIBs. Indeed, the final stage in morphogenesis may occur concomitantly with the release of virus from the cell. The ability of BTV to replicate in and be released from cells may have evolved because of the requirement for insect transmission of the virus.

Studies on the addition of lysomotropic agents suggest that the kinetics of BTV replication depend on superinfection of cells with progeny virions. The difference in sensitivity of reovirus and BTV to lysomotropic agents added at different time p.i. may be due to the length of time developing virus particles retain their ability to generate mRNA. The addition of at least some of the BTV outer coat proteins at the VIB periphery may block transcription at a relatively early-stage of virus particle morphogenesis compared with reovirus where inhibition of transcription by σ-3 occurs at a late stage in virus morphogenesis. In BTV-infected cells a shorter length of time during which particles are active in transcription, perhaps coupled with a decreased rate of mRNA production compared with reovirus (Van Dijk and Huismans 1980), may generate only low levels of mRNA with a consequent delay in the amplification process. Superinfection of BTV-infected cells by progeny virions effectively increase the multiplicity of infection, thereby enhancing the kinetics of virus replication.

Little is known of the mechanism whereby VP2 and VP5 are added to developing virus particles. The initial event in the process appears to occur at the VIB periphery where both fibrillar NS1 and some VP2 become associated with core-like particles. The presence of NS1 in purified core particles suggests that the addition of NS1 may precede the addition of VP2. The absence of large numbers of virus particles close to VIBs late in infection suggests that these putative incomplete particles are released into the cytosol and it is there or probably after binding to the intermediate filaments that more VP2 is added.

NS1 may be removed from the virus particles either concomitantly with the addition of VP2 or as a separate and unrelated event. Some NS1 remains associated, however, with the core particle. The data presented raise the possibility that tubules are formed by condensation of NS1-containing fibrils released from virus-like particles at this stage of morphogenesis. The presence of VP3 and VP7 in tubules may be due to NS1-mediated removal of these proteins from the core-like particle.

The size and heterogeneity of cytoskeleton-associated virus-like particles suggests that VP2 (and VP5) may be added to developing virus particles in that location. If this is the case, the different size of released and cytoskeletal-associated virus particles indicates that either excess VP2 (and VP5) may be removed from cytoskeletal virus or the latter particles undergo a morphological rearrangement of the outer coat prior to release from the cell. The presence of a smaller number of gold-labeled VP2 antibodies on released virus compared with the cytoskeletal particles is consistent with either of these alternatives. It seems likely that cytoskeletal-associated virus particles may reach the cortical layer underlying the cell membrane via the cytosol. Certainly the similar size of released viruses and viruses in the cytosol distinguishes them from cytoskeletal-associated particles. However, cytosolic viruses can themselves be distinguished from released particles by virtue of their ability to bind more gold-labeled anti-VP2 antibody. This suggests that a further loss of VP2 or morphological rearrangement may occur either prior to or concomitant with "budding" or penetration of the virus through the plasma membrane of infected cells.

Acknowledgments. We would like to thank Terry Wise, Gary Crameri, and Andrea Conolan for technical assistance, Dr Allan Gould for the helpful discussions, and John White for supplying all the monoclonal antibodies used in our studies. We would also like to acknowledge the contributions of Dr John Martyn, Helen Wood, and Melinda Cairns in the preparation of the manuscript.

References

Acs G, Klett H, Schonberg M, Christman J, Levin DH, Silverstein SC (1971) Mechanism of reovirus double-stranded ribonucleic acid synthesis in vivo and in vitro. J Virol 8: 684–689

Astell C, Silverstein SC, Levin DH, Acs G (1972) Regulation of the reovirus RNA transcriptase by a viral capsomere protein. Virology 48: 648–654

Babiss LE, Luftig RB, Weatherbee JA, Weihing RR, Ray UR, Fields BN (1979) Reovirus serotypes 1 and 3 differ in their in vitro association with microtubules. J Virol 30: 863–874

Borden EC, Shope RE, Murphy FA (1971) Physiochemical and morphological relationships of some arthropod-borne viruses to bluetongue virus—a new taxonomic group. Physiochemical and serological studies. J Gen Virol 13: 261–271

Borsa J, Morash BD, Sargent MD, Copps TP, Lievaart PA, Szekely JG (1979) Two modes of entry of reovirus particles into L cells. J Gen Virol 45: 161–170
Bowne JG, Jochim MM (1967) Cytopathologic changes and development of inclusion bodies in cultured cells infected with bluetongue virus. Am J Vet Res 28: 1091–1105
Bowne JG, Ritchie AE (1970) Some morphological features of bluetongue virus. Virology 40: 903–911
Brown DT, Burlingham BT (1973) Penetration of host cell membranes by adenovirus 2. J Virol 12: 386–396
Cromack AS, Blue JL, Gratzek JB (1971) A quantitative ultrastructural study of the development of bluetongue virus in Madin–Darby bovine kidney cells. J Gen Virol 13: 229–244
Cowley JA, Gorman BM (1987) Genetic reassortants for identification of the genome segment coding for the bluetongue virus hemagglutinin. J Virol 61: 2304–2306
Dales S (1973) Early events in cell-animal virus interactions. Bacteriol Rev. 37: 103–135
De Sena J, Torian B (1980) Studies on the in vitro uncoating of poliovirus. III. Roles of membrane-modifying and -stabilizing factors in the generation of subviral particles. Virology 104: 149–163
Eaton BT, Crameri GS (1989) The site of bluetongue virus attachment to glycophorins from a number of animal erythrocytes, J Gen Virol (in press)
Eaton BT, Hyatt AD (1989) Association of bluetongue virus with the cytoskeleton. Subcell Biochem 15: 229–269
Eaton BT, Hyatt AD, White JR (1987) Association of bluetongue virus with the cytoskeleton. Virology 157: 107–116
Eaton BT, Hyatt AD, White JR (1988) Localization of the nonstructural protein NS1 in bluetongue virus-infected cells and its presence in virus particles. Virology 163: 527–537
Els HJ, Verwoerd DW (1969) Morphology of bluetongue virus. Virology 38: 213–219
Fenwick ML, Cooper PD (1962) Early interactions between poliovirus and ERK cells: some observations on the nature and significance of rejected particles. Virology 18: 212–223
Foster NM, Alders MA (1979) Bluetongue virus: a membraned structure. Proc Annu Meet Electron Microsc Soc Am 37: 48–49
Furwichi Y, Shatkin AJ (1977) 5′-Termini of reovirus mRNA: ability of viral cores to form caps post-transcriptionally. Virology 77: 566–578
Gould AR (1987a) The complete nucleotide sequence of bluetongue virus serotype 1 RNA 3 and a comparison with other geographic serotypes from Australia, South Africa and the United States of America, and with other orbivirus isolates. Virus Res 7: 169–183
Gould AR (1987b) Conserved and nonconserved regions of the outer coat protein, VP2, of the Australian bluetongue serotype 1 virus, revealed by sequence comparison to the VP2 of North American BTV serotype 10. Virus Res 9: 145–158
Gould AR (1988) Nucleotide sequence of the Australian bluetongue virus serotype 1 RNA segment 10. J Gen Virol 69: 945–949
Gould AR, Pritchard LI (1988) The complete nucleotide sequence of the outer coat protein, VP5, of the Australian bluetongue virus (BTV) serotype 1 reveals conserved and non-conserved sequences. Virus Res 9: 285–292
Gould AR, Hyatt AD, Eaton BT (1988a) Morphogenesis of a bluetongue virus variant with an amino acid alteration at a neutralization site in the outer coat protein, VP2. Virology 165: 23–32
Gould AR, Pritchard LI, Tavaria MD (1988b) Nucleotide and deduced amino acid sequences of the non-structural protein, NS1, of Australian and South African bluetongue virus serotype 1. Virus Res 11: 97–107
Grubman MJ, Appleton JA, Letchworth GL (1983) Identification of bluetongue virus type 17 genome segments coding for polypeptides associated with virus neutralization and intergroup reactivity. Virology 131: 355–366
Hartwig JH, Niederman R, Lind SE (1985) Cortical actin structures and their relationship to mammalian cell movements. Subcell Biochem 11: 1–49
Helenius A, Mellman I, Wall D, Hubbard A (1983) Endosomes. Trends Biochem Sci 8: 245–250
Howell PG, Verwoerd DW (1971) Bluetongue virus. In: Hess WR, Howell PG, Verwoerd DW (eds) African swine fever virus, bluetongue virus. Springer, Berlin Heidelberg pp 35–74 (Virology monographs, vol 9)
Howell PG, Verwoerd DW, Oellermann RA (1967) Plaque formation by bluetongue virus. Onderstepoort J Vet Res 34: 317–332
Huismans H (1970) Macromolecular synthesis in bluetongue virus infected cells. II. Host cell metabolism. Onderstepoort J Vet Res 37: 199–210

Huismans H, Els HJ (1979) Characterization of the tubules associated with the replication of three different orbiviruses. Virology 92: 397–406

Huismans H, Joklik WK (1976) Reovirus-coded polypeptides in infected cells: isolation of two native monomeric polypeptides with affinity for single-stranded and double-stranded RNA, respectively. Virology 70: 411–424

Huismans H, Van der Walt NT, Cloete M, Erasmus BJ (1983) The biochemical and immunological characterization of bluetongue virus outer capsid polypeptides. In: Compans RW, Bishop DHL (eds) Double-stranded RNA viruses. Elsevier, New York

Huismans H, Van der Walt NT, Cloete M, Erasmus BJ (1987a) Isolation of a capsid protein of bluetongue virus that induces a protective immune response in sheep. Virology 157: 172–179

Huismans H, Van Dijk AA, Bauskin AR (1989b) In vitro phosphorylation and purification of a nonstructural protein of bluetongue virus with affinity for single stranded RNA, J Virol 61: 3589–3595

Huismans H, Van Dijk AA, Els HJ (1987c) Uncoating of parental bluetongue virus to core and subcore particles in infected L cells. Virology 157: 180–188

Hyatt AD, Eaton BT (1988) Ultrastructural distribution of the major capsid proteins within bluetongue virus and infected cells. J Gen Virol 69: 805–815

Hyatt AD, Eaton BT, Lunt R (1987) The grid-cell-culture-technique: the direct examination of virus-infected cells and progeny viruses. J Microsc 145: 97–106

Hyatt AD, Eaton BT, Brookes SB (1989) The release of bluetongue virus from infected cells and their superinfection by progeny virus. Virology 173: 21–34

Joklik WK (1981) Structure and function of the reovirus genome. Microbiol Rev 45: 483–501

Kaljot KT, Shaw RD, Rubin DH, Greenberg HB (1988) Infectious rotavirus enters cells by direct cell membrane penetration, not by endocytosis. J Virol 62: 1136–1144

Lecatsas G (1968) Electron microscopic study of the formation of bluetongue virus. Onderstepoort J Vet Res 35: 139–149

Mackow ER, Shaw RD, Matsui SM, Vo PT, Dang MN, Greenberg HB (1988) The rhesus rotavirus gene encoding protein VP3: location of amino acids involved in homologous and heterologous rotavirus neutralization and identification of a putative fusion region. Proc Natl Acad Sci USA 85: 645–649

McPhee DA, Parsonson IM, Della-Porta AJ (1982) Comparative studies on the growth of Australian bluetongue virus serotypes in continuous cell lines and embryonated chicken eggs. Vet Microbiol 7: 401–410

Mertens PPC, Brown F, Sangar DV (1984) Assignment of the genome segments of bluetongue virus type 1 to the proteins which they encode. Virology 135: 207–217

Mertens PPC, Burroughs JN, Anderson J (1987) Purification and properties of virus particles, infectious subviral particles, and cores of bluetongue virus serotypes 1 and 4. Virology 157: 375–386

Morgan EM, Zweerink HJ (1975) Characterization of transcriptase and replicase particles isolated from reovirus-infected cells. Virology 68: 455–466

Ogier G, Chardonnet Y, Doerfler W (1977) The fate of type 7 adenovirions in lysosomes of HeLa cells. Virology 77: 67–77

Ohkuma S, Poole B (1978) Fluorescence probe measurement of the intralysosomal pH in living cells and the perturbation of pH by various agents. Proc Natl Acad Sci USA 75: 3327–3331

Owen NC, Munz EK (1966) Observations on a strain of bluetongue virus by electron microscopy. Onderstepoort J Vet Res 33: 9–14

Sharpe AH, Chen LB, Fields BN (1982) The interaction of mammalian reoviruses with the cytoskeleton of monkey kidney CV-1 cells. Virology 120: 399–411

Skup D, Millward S (1980) mRNA capping enzymes are masked in reovirus progeny subviral particles. J Virol 34: 490–496

Sturzenbecker LJ, Nibert M, Furlong D, Fields BN (1987) Intracellular digestion of reovirus particles requires a low pH and is an essential step in the viral infectious cycle. J Virol 61: 2351–2361

Tyler KL, Fields BN (1985) Reovirus and its replication. In: Fields BN (ed) Virology. Raven, New York

Van Dijk AA, Huismans H (1980) The in vitro activation and further characterization of the bluetongue virus-associated transcriptase. Virology 104: 347–356

Van Dijk AA, Huismans H (1988) In vitro transcription and translation of bluetongue virus mRNA. J Gen Virol 69: 573–581

Verwoerd DW (1969) Purification and characterization of bluetongue virus. Virology 38: 203–212

Verwoerd DW (1970) Diplorna viruses: newly recognized group of double-stranded RNA viruses. Prog Med Virol 12: 192–210
Verwoerd DW, Huismans H (1972) Studies on the in vitro and the in vivo transcription of the bluetongue virus genome. Onderstepoort J Vet Res 39: 185–191
Verwoerd DW, Louw H, Oellermann RA (1970) Characterization of bluetongue virus ribonucleic acid. J Virol 5: 1–7
Verwoerd DW, Els HJ, De Villiers EM, Huismans H (1972) Structures of the bluetongue virus capsid. J Virol 10: 783–794
Verwoerd DW, Huismans H, Erasmus BJ (1979) Orbiviruses. In: Fraenkel-Conrat H, Wagner RR (eds) Comprehensive virology, vol. 14. Plenum, New York
Watanabe Y, Millward S, Graham AF (1968) Regulation of transcription of the reovirus genome. J Mol Biol 36: 107–123
Zarbl H, Millward S (1983) The reovirus multiplication cycle. In: Joklik WK (ed) The Reoviridae. Plenum, Yew York
Zarbl H, Skup D, Millward S (1980) Reovirus progeny subviral particles synthesize uncapped mRNA. J Virol 34: 497–505
Zweerink HJ, McDowell MJ, Joklik WK (1971) Essential and non-essential noncapsid reovirus proteins. Virology 45: 716–723
Zweerink HJ, Morgan EM, Skyler JS (1976) Reovirus morphogenesis: characterization of subviral particles in infected cells. Virology 73: 442–453

Pathology and Pathogenesis of Bluetongue Infections

I. M. Parsonson

1 Introduction . . . 119

2 Bluetongue Disease in Sheep . . . 120
2.1 Clinical Signs . . . 120
2.2 Pathology of Bluetongue Disease in Sheep . . . 122
2.3 Gross Pathology . . . 122
2.4 Clinical Pathology of Bluetongue Disease in Sheep . . . 123
2.5 Histological Findings in Bluetongue Disease of Sheep . . . 123
2.6 Pathogenesis of Bluetongue Disease in Sheep . . . 124

3 Distribution of Bluetongue Virus in Blood . . . 126

4 Bluetongue Virus Infection in Cattle . . . 128
4.1 Bluetongue Disease in Cattle . . . 128

5 Effect of Bluetongue Virus on Reproduction in Sheep . . . 130

6 Effect of Bluetongue Virus on Reproduction in Cattle . . . 131
6.1 Congenital Bluetongue Virus Infections in Cattle . . . 133

7 Bluetongue Virus Infection in other Species . . . 134
7.1 Goats . . . 134
7.2 Wildlife Species in the USA . . . 134
7.3 Wildlife Species in Africa . . . 134

8 Epizootic Hemorrhagic Disease (EHD) of Deer Virus Infection in Cattle . . . 135
8.1 EHD Virus in Australia . . . 135
8.2 EHD Virus in Japan . . . 135
8.3 EHD Virus in the United States . . . 135

9 Conclusions . . . 136

References . . . 136

1 Introduction

Bluetongue disease is an infectious, noncontagious, arthropod-borne viral disease, mostly of sheep, but also of other ruminants (Erasmus 1975). The original detailed descriptions of bluetongue disease in sheep were those of Spreull (1905).

The identification of bluetongue virus (BTV) in a number of countries of the Middle East, Asia, and the USA in the early 1940s and 1950s led to the description of bluetongue as an emerging disease (Howell 1963). BTVs are now known to occur in tropical, semitropical, and temperate zones of the world within

Ocean Grove 3226, Victoria, Australia

an equatorial belt which extends approximately between the latitudes 40° N and 35° S (Shimshony et al. 1988).

An important factor in the distribution of BTV worldwide is the availability of suitable vectors, usually biting midges of the species *Culicoides* (Wirth and Dyce 1985). Wherever the required vectors are present, BTV can become endemic and infected vectors can often be transported by prevailing winds to areas, where if they come into contact with susceptible animals, they may infect them thereby instituting epizootics. The wind-borne introduction of BTV to various countries of the Mediterranean littoral has been postulated (Sellers 1975; Taylor et al. 1985; Polydorou 1985). An epizootic of BTV-10 infection occurred on the Iberian peninsula (Silva 1956; Manso-Ribeiro et al. 1957). Epizootics due to BTV infection have occurred in Australia and the Okanaghan Valley of Canada as extensions from endemic areas (Murray 1987; Thomas et al. 1982).

Monitoring the prevalence of serum antibodies to BTV in sentinel cattle, combined with insect trapping, has enabled the determination of the ecological range of *Culicoides brevitarsis*, one of the known vectors of BTV in Australia (Muller et al. 1982; Murray 1987).

2 Bluetongue Disease in Sheep

Bluetongue is primarily a disease of sheep and most breeds of sheep are susceptible, although not to the same degree. In many areas of the world where BTV is endemic the indigenous sheep are less susceptible than the exogenous breeds such as Merinos and European mutton types (Erasmus 1975; Barzilai and Shimshony 1985; Taylor et al. 1985; Hafez and Taylor 1985; Gibbs and Greiner 1985; Tamayo et al. 1985; St. George 1985).

In the first reports of bluetongue disease in the USA (Hardy and Price 1952) it was noted that nursing lambs were less susceptible than older animals. Luedke et al. (1964) confirmed this observation under experimental conditions and noted that susceptibility to infection varied considerably among individuals of a breed with older sheep being more susceptible than lambs. However, in other reports no age differences were found (Erasmus 1975).

In Australia although eight different serotypes of BTV have been identified, no clinical disease has been reported in sheep under natural grazing conditions (St. George and McCaughan 1979; Uren and Squire 1982; Uren and St. George 1985; Flanagan et al. 1982).

2.1 Clinical Signs

Most cases of BTV infection of sheep are subclinical or mild and prognosis is generally favorable with complete recovery within a few weeks. Erasmus (1975) has described a mortality rate varying from 2% to 30% under field conditions in

South Africa. In the USA, morbidity rates of less than 30% and averaging 10% with a mortality rate estimated to be 5% have been reported from field outbreaks (HARDY and PRICE 1952; MCKERCHER et al. 1953; MOULTON 1961).

The clinical signs of bluetongue disease following natural infection and the gross pathology resulting were described in South Africa by SPREULL (1905), THEILER (1906), DIXON (1909), and THOMAS and NEITZ (1947); in Cyprus by GAMBLES (1949); and in the United States by HARDY and PRICE (1952) and MOULTON (1961).

There is a great variability in the response of sheep to BTV infection which can range from inapparent to acute and fulminating. The severity of the disease depends on the susceptibility of the animals, the pathogenicity of the virus, and the environmental conditions. Sheep that are stressed severely or subjected to high solar radiation levels were reputed to develop more severe lesions than those not so treated (NEITZ and RIEMERSCHMID 1944).

The descriptions of the clinical signs and lesions in natural infections are generally in agreement in all reports and are as follows. The first sign is a rise in body temperature which may be transitory or continue for up to 14 days, with the average duration about 5 to 7 days (ERASMUS 1975). An increased respiratory rate which may be moderate, or hyperpneic in the more acute cases, occurs early in the febrile response. Hyperemia and swelling of the buccal and nasal mucosa are often observed at the time of the first temperature rise. The swelling is usually accompanied by frothy salivation, swollen tongue, and licking of the lips and nostrils. This is followed by cracking of the epidermis at the commisures of the lips, with encrustation of these areas and the nasolabial plane combined with edema of the face, submaxillary space, and on occasion, the ears and eyelids. At this time hemorrhages ranging from petechiae to ecchymotic occur in the mucous membrane of the gingiva and oral cavity in the vicinity of the ventral surface of the tongue. Very rarely the tongue becomes swollen, markedly congested, and may protrude from the mouth and because of its cyanotic appearance gives the name "bluetongue" to the disease. As this sign is transitory it is often missed.

The epithelial lesions of the mouth and nasolabial plane may become chronic. The thickened epithelium results in excoriation, leaving bleeding, ulcerated lesions which become infected and often necrotic.

During this time acutely infected sheep develop stiffness, lethargy, and anorexia. The reluctance to move has been related to the appearance of reddening around the coronet of the hoof, and the development of petechial hemorrhages which occur at the skin horn junction (periople). Later the breakdown products of these hemorrhages underrun the horn as red streaks in white-hoofed animals. In addition to the hoof lesions moderate to severe skeletal muscle damage occurs leading to degeneration of the muscles, extreme weakness, and prostration. Torticollis may result and this is usually terminal. Occasionally, sheep will vomit as a result of smooth muscle lesions of the esophagus and pharyngeal area. When vomiting occurs aspiration of ruminal contents and pneumonia inevitably follows, ending frequently in death.

2.2 Pathology of Bluetongue Disease in Sheep

There are a number of excellent descriptions of the gross pathological lesions in sheep at necropsy (Bekker et al. 1934; Thomas and Neitz 1947; Moulton 1961; Luedke et al. 1964; Parish et al. 1982) which between them cover the majority of findings.

2.3 Gross Pathology

The gross lesions are dependent on the stage of the disease at which the sheep dies or is killed, and on the particular infecting virus serotype and strain. In addition, environmental effects can also play a role.

The external lesions are generally visible during the latter stage of the disease. Edema, which is a feature of the head and face, usually extends into the subcutaneous pendulous areas of the neck and is present in the trachea and larynx. Edematous fluid is often present in the submandibular space and in ventral areas of the neck and thorax. Subcutaneous edema can also extend to the ventral abdomen and intrafascicular muscle planes of the fore- and hindlimbs. In the majority of animals the subcutaneous edema is not extensive. However, in animals that die in the acute stages of bluetongue disease, the pleural, pericardial, and peritoneal cavities usually contain excessive amounts of yellow serous fluid. The skeletal muscle lesions can be seen as hemorrhages or hemoglobin-stained gelatinous areas among muscle fibers and may be scattered throughout several muscles in a group (Thomas and Neitz 1947; Moulton 1961).

Lesions of the digestive tract include lesions of the oral cavity with extension of edema into the laryngeal and pharangeal area. Damage to the esophageal musculature has been reported (Mahrt and Osburn 1986) and has been suggested as a possible cause of the vomiting sometimes seen prior to the development of aspiration pneumonia (Luedke et al. 1964; Erasmus 1975). Ecchymotic hemorrhages may be found under the peritoneal lining of the rumen and reticulum. Hyperemia of the ruminal papillae, pillars, and reticular folds may also be present (Erasmus 1975), as well as congestion and hemorrhages in the region of the abomasum and duodenum (Mahrt and Osburn 1986).

Hemorrhages and congestion have also been reported for the cardiovascular system particularly the pulmonary artery, subendocardial areas, and papillary muscles of the heart (Erasmus 1975; Uren and Squire 1982; Mahrt and Osburn 1986).

Lesions of the respiratory system often begin in the trachea with petechial hemorrhages in the mucous membrane and froth in the lumen extending into the bronchi. In chronic cases of long duration, the ventral lobes of the lungs may have areas of atelectasis, or if vomiting has occurred, bronchopneumonia may be present with evidence of ruminal material in the trachea and bronchi (Moulton 1961; Erasmus 1975; Luedke et al. 1964). Lymph nodes of the head and neck are usually congested and enlarged and petechial hemorrhages are commonly present (Moulton 1961; Mahrt and Osburn 1986).

The spleen may be congested, and scattered petechial hemorrhages may be present beneath the capsule. Similar petechiae have been reported in the thymus (LUEDKE et al. 1964).

The kidneys are often congested. Petechial hemorrhages may be seen in the mucosa of the urinary bladder, urethra, and vulva or penile sheath. Bluetongue infection of the pregnant ewe may result in infection of the fetus depending on the stage of gestation. Encephalopathies occurred in fetuses following vaccination of ewes pregnant at 5 to 6 weeks with egg-adapted-BTV serotype 10 vaccines (SHULTZ and DELAY 1955; CORDY and SHULTZ 1961; YOUNG and CORDY 1964; RICHARDS and CORDY 1967).

Lesions seen at necropsy relate to the severity of damage to the microvascular vessels which in turn results in vascular permeability with edema, hemorrhage, thrombosis, ischemia, and necrosis of a wide range of tissues.

2.4 Clinical Pathology of Bluetongue Disease in Sheep

Several studies on the clinical pathology of bluetongue disease have been made (GRAF 1933; LUEDKE et al. 1964; JEGGO et al. 1986). Hematological studies showed that the sheep had leukopenia with lowest counts on days 5 to 7 after infection. Of the sheep examined, neutropenia occurred in 88%, lymphopenia in 95%, and eosinopenia in 77%. Hemolytic anemia was identified by packed cell volume, hemoglobin, and icterus index (LUEDKE et al. 1964). The levels of several plasma enzymes have been determined and glutamic oxalacetic transaminase, glutamic pyruvic transaminase, lactic dehydrogenase, and aldose may increase and reach a peak about 8 days after the peak of pyrexia. Significant changes were found in serum enzyme activities in creatinine kinase (JEGGO et al. 1986; MAHRT and OSBURN 1986), aldolase, and lactate dehydrogenase—all having raised levels from day 10 after infection in the animals that subsequently died. Changes in aspartate amino-transferase levels in plasma occurred in sheep that died and also in sheep that survived and were not correlated to the severity of clinical findings (JEGGO et al. 1986).

2.5 Histological Findings in Bluetongue Disease of Sheep

Microscopic lesions were first described in the stratified squamous epithelium of the skin and oral mucosa of sheep with bluetongue disease by BEKKER et al. (1934). Subsequent description of the microscopic findings by THOMAS and NEITZ (1947) and MOULTON (1961) have expanded the original descriptions. Lesions in acute infections are the result of vascular thrombosis occurring in the capillaries and small blood vessels causing hemorrhage and edema within the surrounding tissues. In the squamous layers of the skin and mucosa, vacuolation and necrosis of the epithelium results. Neutrophilic migration occurred in many areas of the epithelium and mononuclear and polymorphonuclear cells accumulated in the underlying perivascular areas of the dermis. In more chronic lesions, erosions occurred often developing into ulcerated areas. Skeletal muscle degeneration was seen as focal areas within muscle masses, occasionally as degenerated single or several fibers only, and was characterized by swelling, hyaline degeneration of the fibers, and infiltration of neutrophils and fibroblasts. Hemorrhages and edema were also frequently present. Lesions in the myocardium were similar to those seen in skeletal muscle. In sheep with hemorrhages in the tunica media of the pulmonary artery, areas between muscle fibers were infiltrated with extravasated blood, but there was no degeneration of muscle fibers.

Microscopic lesions of pulmonary edema and congestion with atelectasis of areas in the ventral lobes were common. In addition, sheep that had aspirated ruminal material had severe bronchopneumonia, which, depending on the stage, was either diffuse and suppurating or undergoing marked necrosis. The latter finding was frequently associated with pharyngeal or esophageal myodegeneration and hemorrhage.

Other microscopic findings in lymph nodes, and associated with hemorrhages of the smooth muscle areas of the rumen, abomasum, and small intestine are commonly associated with damage to the microvasculature of those areas. Laminitis may occur in some sheep and is characterized microscopically by hemorrhages and edema of the tissues at the skin-horn junction.

Although lesions have been described for other organs and tissues of sheep that have acute or chronic bluetongue disease, the common findings relate to a breakdown in the microvasculature due to thrombosis and subsequent hemorrhage, edema, ischemia, and necrosis of the surrounding tissues.

2.6 Pathogenesis of Bluetongue Disease in Sheep

Erasmus (1975) drew attention to the observations of Bekker et al. (1934) who noted the regular development of lesions in those tissues where mechanical stress was common. They cited the oral cavity in the vicinity of the teeth as well as the muscular pillars and the esophageal groove of the rumen. Thomas and Neitz (1947) confirmed these findings and they considered that vascular lesions were the primary injury. However, neither ventured to explain the mechanisms by which the virus caused the lesions.

Stair (1968) studied the pathogenesis of bluetongue disease in sheep using clinical and gross observations, immunofluorescence, histological, and virus neutralization techniques. The sheep were inoculated intravenously and one or two sheep were killed at daily intervals for 16 days after inoculation. He found that BTV had an affinity for endothelial cells, periendothelial cells, and pericytes of capillaries and small blood vessels. Immunofluorescence was concentrated in the small vessels underlying stratified squamous epithelium and, in particular, that of the oral cavity skin around the mouth and nasolabial plane and the coronet of the hoof. The concentration of viral antigen in these areas was attributed to the lower temperature of these structures in relation to the rest of the body. Basic histopathological lesions of the endothelial cells resulted in nuclear fragmentation and cell death, vacuolation of the cytoplasm, and cytoplasmic swelling. The cellular changes commenced from 6 to 10 days after inoculation. Ischemic lesions in the epithelium followed the primary viral inflammatory changes in the endothelial cells. Mechanical stress and microbiological infection greatly influenced the degree of secondary changes. There was a marked increase in virus neutralizing antibodies from day 6 to day 8 after inoculation. Fever, the initial clinical signs, primary histological lesions, and maximal concentration of viral antigen occurred at the same time that the marked increase in serum neutralizing antibodies commenced. STAIR (1968) speculated that the lesions resulting from BTV infection were generally confined to the areas of lower body temperature and that the vascular lesions and exudation of serum crusts in nonwool areas resulted in a photosensitization-like hypersensitivity reaction in some sheep.

PINI (1976) studied BTV infection in yearling Merino sheep inoculated subcutaneously into the auricula of the ear with plaque purified BTV-10. Virus was isolated from tissue samples taken from individual animals killed at daily intervals over a period of 11 days. He found the incubation period was between 6 and 9 days and that the first clinical sign was pyrexia. The virus was first detected in the draining lymph nodes of the head and cervical region also in tonsils and spleen. Viremia was first detected on day 6 postinoculation. Macroscopic lesions of the oral cavity, lips, and edematous swellings of the face were seen on day 8. From these findings PINI (1976) postulated that BTV entered the regional lymph nodes from infection sites and was deceminated via lymph or vascular systems to other lymphoid tissues where further replication occurred. Following replication, the virus was carried via the vascular system to other organs and predilection sites such as the exposed areas of the body. The fact that viremia was not detected until 48 h after the virus detected in lymphoid tissue is very important as virus replication was possibly occurring in these sites in the preferred cells before release into the bloodstream. The antibody response was detected by immunofluorescence and coincided with a decrease in virus in the blood without appearing to affect the antigen concentration in the tissues. PINI (1976) found no evidence of viral antigen in the tissues of sheep killed at 6, 8, and 16 weeks after infection. Both these studies (STAIR 1968; PINI 1976) confirm the distribution of virus in the tissues and vascular system. Both noted the presence of neutralizing antibody and virus in the bloodstream at the same time.

ERASMUS (1975), in comparing the above studies, considered that the route of infection may have been an important factor in the early distribution of BTV in the vascular endothelium (intravenous, STAIR 1968) and in the lymphoid tissues (intradermal, PINI 1976) in which primary viral replication occurred. Natural transmission of BTV by *Culicoides sp.* is probably via intradermal routes with virus replication in draining lymphoid tissues before general release of virus into the bloodstream.

In a study of the pathogenesis of BTV-4 (Cyprus strain) in sheep, LAWMAN (1979) found that the primary replication site of the virus was in the lymphoreticular system. Once viremia was detected the virus could be identified in a number of tissues and organs. He found the major hematological feature of BTV infection in sheep, goats, and cattle to be a transient pan-leukopenia with maximum leukopenia preceding both the peaks of viremia and pyrexia. In this study the cells involved in primary replication were monocytes and macrophages, although replication also occurred in endothelial cells and neutrophils.

3 Distribution of Bluetongue Virus in Blood

There are a number of reports describing the distribution of BTV in the various components of the blood. SPREULL (1905) found that serum was as infective as whole blood and concluded that the microbe was not bound up in red corpuscles and suggested dropping the term "malarial catarrhal fever," and instead using bluetongue as the name for the disease. PINI et al. (1966) investigated titration of BTV from plasma and buffy coat fractions of sheep blood and concluded that virus was associated with the buffy coat.

LUEDKE (1970) in experiments in sheep, goats, and cattle found that concentration of BTV in the erythrocyte fraction of blood was 10 to 100 times that of the virus in buffy coat fractions and concluded that BTV was closely associated with the erythrocytes during all stages of infection. SAWYER and OSBURN (1977) noted that BTV can be isolated with equal efficiency from mononuclear and erythrocyte-granulocyute fractions of blood. PARSONSON et al. (1987a) found no differences in the isolation of BTV-20 in cattle from crude buffy coat or erythrocyte fractions using several tissue culture cell lines and embryonating chicken eggs. MORRILL and MCCONNELL (1985) examined cellular elements from the blood of calves experimentally inoculated with BTV-10, BTV-11, BTV-13, and BTV-17 using transmission electron microscopy. They found intravacuolar viral particles within infected agranular leukocytes from blood samples collected on day 14 post inoculation. No virus particles were seen associated with other cellular elements including erythrocytes and platelets.

Although the literature in regard to this aspect is confusing, there can be no perfect separation of blood fractions without some contamination; thus, the discussion becomes academic. Based on the evidence for replication of BTV in cells where the virus is closely adherent and associated with cytoplasmic cellular structures nucleated cells are necessary (HOWELL et al. 1967; LAWMAN 1979; MCPHEE et al. 1982). Unfortunately there is no definitive information available on the cell surface receptor sites to which BTV may bind. In an in vitro study of BTV-erythrocyte combinations there were indications that the virus binds to specific sialic acid-containing sites in human glycophorins and a number of animal erythrocytes (EATON and CRAMERI 1989). However, because of the close association of BTV with the cell surface, once the virus has passed through the cell membrane (see Chap. 4, this volume), and because of the affinity of the virus binding to specific sialic acid sites, shed complete virions could adhere to, and be transported by, erythrocytes as well as the cells of the vascular system in which they may replicate. Obviously the methods used for separating the virus from the cell membranes of erythrocytes and buffy coat cells are of importance in establishing infection and determining the titers of viremias in animals.

In epidemiological studies of the prevalence of antibodies to BTV in endemic areas of Australia, St. George (1985) found antibodies to BTV and EHD viruses in cattle, buffaloes, deer, goats, and sheep. No antibodies have been found in pigs, horses, marsupials (kangaroos and wallabies), or humans. It would appear that there must be some specific factors in the virus-host relationship associated with virus accessibility to host cells or a specific arthropod-vector-host relationship with respect to the transmission of BTV.

Specific BTV neutralizing antibodies are generally detected from 7 to 14 days after infection in sheep and are maintained for extended periods (Howell 1960; Luedke and Jochim 1968; Luedke 1970; Pini 1976) often coexisting in the blood with the specific BTV. Erasmus (1975) suggested that this may be due to the ability of the virus to become sequestrated in erythrocyte membranes and thereby avoid neutralization. Even very low levels of serotype-specific (homotypic) neutralizing antibodies present at the time of the challenge to immunity appear to afford protection against heterotypic BTV challenge (Groocock et al. 1982; Jeggo et al. 1983; Parsonson and Luedke, unpublished).

Huismans and Erasmus (1981) found that BTV outer-capsid polypeptide P2 appeared to be the main determinant of serotype specificity, whereas the main determinant of group specificity resided in the core protein P7. Monoclonal antibodies raised against P2 were able to neutralize BTV-17 (Appleton and Letchworth 1983) and provide passive immunity and protection against challenge with a pathogenic strain of the homologous serotype (Letchworth and Appleton 1983). This was confirmed by Huismans et al. (1987) with the specific purified antigen P2 which fully protected sheep against challenge with the homologous BTV serotype (BTV-10). In a comparison with BTV-10 (Richards et al. 1988), the appearance of neutralizing antibody in serum did not coincide with the clearance of virus from blood and specific neutralizing antibody and virus coexisted in peripheral blood for as long as 28 days. It was concluded that the intensity of the humoral immune response to specific BTV proteins may influence the duration of viremia. The importance of VP2 in the ability of BTV-10 to bind to BHK cells was shown by Huismans et al. (1983). Similarly, Cowley and Gorman (1987) using reassortants of BTV-20 and BTV-21 showed that agglutination of erythrocytes (BTV-20 sheep only, and BTV-21 sheep, bovine, human, and goose), correlated with the presence of VP2.

The coexistence of BTV and neutralizing antibodies in the peripheral circulation during viremia has often been cited (Stair 1968; Luedke 1969, 1970; Pini 1976; Groocock et al. 1982; MacLachlan and Fuller 1986), and virus clearance in calves infected with BTV-10 was shown to coincide with humoral immune responses to VP2 and nonstructural proteins NS1 and NS2 (MacLachlan et al. 1987).

Suggestions have been made that BTV may reside within circulating cell types and remain protected from the effects of neutralizing antibody (LUEDKE 1970; MACLACHLAN et al. 1987), or that antigen variation of the virus as a result of immunological pressure by neutralizing antibody may result in antigenic drift. Further possibilities are that BTV released from cells in which replication has occurred may differ from infectious particles that are present in the cell cytoplasm or associated with the cytoskeleton and which may be released on cell death and lysis (EATON, HYATT and BROOKES, Chap. 4). Subtle differences in the conformation of BTV outer-coat proteins may result in lowering neutralizing antibody avidity and also allow the possibility of neutralizing activity to heterologous BTV which have often been puzzling features of bluetongue disease. DELLA-PORTA et al. (1983) found positive serum neutralizing reactions of low titer to a number of different BTV group viruses even with a single serum and this pattern was particularly evident in animals from the BTV endemic areas of Northern Australia. In South Africa, OWEN et al. (1965) described the isolation of a number of heterologous BTV types from individual cattle over a relatively short time indicating repeated reinfection. These workers suggested that no basic immunity to the various BTV types was present in cattle. DAVIES (1978) reported on the possibility of a number of BTV serotypes infecting individual animals in the one area in 1 year in Kenya during which time 19 BTV serotypes were identified.

Antigenic drift does not appear to be the mechanism for the prolonged viremia that occurs in some cattle (HEIDNER et al. 1988). The possibility of sequestered virus or differences between released, cytoplasmic, and cytoskeletal virus remain to be investigated.

4 Bluetongue Virus Infection in Cattle

BTV infections of cattle have been recognized worldwide and the epizootiology of BTV has been reviewed in Africa (ERASMUS 1975; DAVIES 1980), Australia (DELLA-PORTA et al. 1983), and in the USA (HOURRIGAN and KLINGSPORN 1975; METCALF et al. 1981). Infections in cattle are now known to occur in many areas of the world (OZAWA 1985). However, descriptions of overt disease in cattle due to BTV are very rare and generally apply to small numbers of animals during epizootics or in isolated episodes (BEKKER et al. 1934; KOMAROV and GOLDSMIT 1951; SILVA 1956; LOPEZ and SANCHEZ BOTIJA 1958; BOWNE et al. 1968; LUEDKE et al. 1970).

4.1 Bluetongue Disease in Cattle

SPREULL (1905) demonstrated that BTV could be present in the peripheral blood of a calf infected 21 days previously. The first report of clinical disease in cattle due to BTV was that of BEKKER et al. (1934), who claimed to reproduce the disease.

De Kock et al. (1937) and Mason and Neitz (1940) attempted to repeat the experimental initiation of a clinical disease in cattle using inocula as prepared by Bekker et al. (1934), and were able to isolate BTV from cattle exposed to natural infection but were unable to reproduce the clinical disease described by Bekker et al. (1934). Their conclusions were that the disease described was ulcerative stomatitis and that BTV only caused a subclinical infection in cattle. These findings are very interesting and important because the report by Bekker et al. (1934) has been referred to frequently in the literature but the reports of De Kock et al. (1937) and Mason and Neitz (1940), refuting the earlier results, are rarely quoted. Particularly as Owen et al. (1965) pointed out, "Cattle in South Africa can harbour a number of antigenically different types of bluetongue virus without showing signs of illness. It, therefore, appears that when investigating suspected bluetongue in cattle, caution must be exercised against the co-incidental isolation of bluetongue virus, and its correlation with the symptomatoloy encountered."

Davis (1978) monitored a herd of 3000 head of cattle under continuous challenge by BTV, and paid particular attention to heifers in their first breeding season. He found that BTV did not result in abortions, stillbirths, or fetal abnormalities of any kind and contrasted this finding with the situation in the USA reported by Luedke et al. (1970). BTV infection of cattle is relatively common in the western, central, and southeastern regions of the USA where serum antibody prevalence to the virus varies from 6% to 60% of animals tested (Metcalf et al. 1981). Isolation of BTV from cattle blood was carried out over a 4-year period in California over the months of August through October. Four to eight % of the cattle were viremic to BTV, but without any evidence of clinical disease (Stott et al. 1982).

Experimental reproduction of clinical bluetongue disease in cattle has never been adequately demonstrated, and the only signs reported are transient mild pyrexia, leukopenia, viremia, and production of group and specific neutralizing antibodies (Mason and Neitz 1940; Owen et al. 1965; Bowne et al. 1968; Luedke et al. 1977; St. George and McCaughan 1979; Groocock et al. 1983; Bowen et al. 1987a, b; MacLachlan et al. 1987).

The only reports of clinical bluetongue disease experimentally produced in cattle are those of Bekker et al. (1934) and Luedke et al. (1977), the latter following bites of infected *Culicoides variipennis*. Stott et al. (1982) using inactivated BTV and immunomodulatory drugs followed by challenge with virulent BTV-17, produced a clinical reaction 76 days after immunization. More recently Anderson et al. (1985) described the development of overt clinical bluetongue disease as appearing to be due to previous sensitization by BTV or BTV-related viruses. This proposal does not seem likely in view of the continuous challenge of BTV and related orbiviruses reported from all endemic areas of the world and the absence of reports of clinical bluetongue disease in cattle.

Other infectious agents may confuse the clinical syndromes reported for cattle. Referring to the early clinical and pathological descriptions of Bekker et al. (1934) and Mason and Neitz (1940), the similarity to pestivirus infection of cattle is striking (Radostits and Littlejohns 1988). From the reports of Owen et al. (1965) and DAVIES (1978) a number of different BTV serotypes can be isolated from sentinel cattle over relatively short periods of time ranging from within weeks to a few months. Isolation of BTV (now designated BTV-4, HOWELL 1970) from the clinical cases described by BEKKER et al. (1934) may have been fortuitous, and although inoculums from infected animals reproduced the disease, the inoculums consisted of tissues, secretions, and various other material and may have contained several infectious agents including BTV. BEKKER et al. (1934) concluded that while cattle were susceptible to BTV they were far more resistant than sheep. They commented on the few clinically infected animals seen only in sporadic isolated field outbreaks. MASON and NEITZ (1980) were not supportive of the findings of BEKKER et al. (1934) and were able to show that erosive stomatitis was not caused by the "Bekker" strain (BTV-4) in cattle.

Clinical disease in cattle has been described by KOMAROV and GOLDSMIT (1951) in Israel during an epizootic of bluetongue disease. However, no clinical disease has been seen in cattle since 1964 (SHIMSHONY et al. 1988) despite evidence of BTV infections occurring regularly. Since then infection has not affected the health status of Israeli–Holstein cows or bulls over a 25-year period.

Sentinel cattle in Kenya that had seroconverted to 19 BTV serotypes were closely observed over a 10-year period for abortions and fetal abnormalities. None found attributable to BTV have occurred (DAVIES 1980).

The cattle population in the northern areas of Australia have a high prevalence of serum antibodies to BTV where these viruses are endemic, but there is no evidence of clinical disease or reproductive problems (PARSONSON and SNOWDON 1985). Occasional epizootics of BTV infection of cattle have been reported in Canada (THOMAS et al. 1982) and in the northwestern United States (PARISH et al. 1982), without evidence of congenital infections or clinical disease in cattle.

A serological survey of ruminants in some Caribbean and South American countries for type-specific antibody to BTV and EHD virus and analysis of the data indicated that in 1981–1982 BTV-6, BTV-14, and BTV-17 or closely related viruses were infecting ruminants and EHD 1 was infecting cattle. It was deduced that an endemic situation is present in these regions without evidence of clinical disease (GUMM et al. 1984).

HERNIMAN et al. (1980), found serological evidence of widespread BTV infections throughout the Eastern Hemisphere in tropical, subtropical, and Mediterranean regions, yet there are no reports of clinical disease in cattle or buffaloes.

5 Effect of Bluetongue Virus on Reproduction in Sheep

LUEDKE (1985), discussing the effect of BTV on reproduction in sheep and cattle, commented that BTV has been shown to be abortigenic and teratogenic following transmission under natural or experimental conditions in sheep and cattle. The major cause of fetal abnormalities or abortions in sheep has been shown to be due to the use of live attenuated BTV vaccines, particularly egg-adapted BTV-10 vaccine. Field outbreaks and experimental studies have resulted in fetal abnormalities, neonatal deaths, or abortions from the use of this virus vaccine in sheep (SHULTZ and DELAY 1955; YOUNG and CORDY 1964; OSBURN et al. 1971a, b; ENRIGHT and OSBURN 1980).

In Israel in a trial using live attenuated BTV (South Africa) vaccine, SHIMSHONY et al. (1980) vaccinated pregnant ewes during the sixth week of pregnancy resulting in fetal loss of 40%; however, the remaining ewes produced progeny which were normal and showed no congenital defects resulting from vaccination.

GIBBS et al. (1979) infected sheep in midgestation with BTV-4 and BTV-6. Normal lambs were born but were viremic for 2 months after birth. It was suggested that this may provide an overwintering mechanism for the virus. BWANGAMI (1978) found that virulent BTV-10 and BTV-11 resulted in death of the fetus rather than in congenital defects. Some evidence of placenta invasion by these BTV serotypes used in the experiment above was found in 30% of the ewes inoculated at the fifth or sixth week of gestation (ANDERSON and JENSEN 1969). When BTV-20 was inoculated into ewes at 35 to 42 days gestation, there was no evidence of transplacental transfer and the lambs were normal at birth. At necropsy 46 days after the birth of the last lamb, no gross or histological lesions were seen (FLANAGAN et al. 1982). Inoculation of BTV-10, BTV-11, BTV-13, and BTV-17 and epizootic hemorrhagic disease (EHD) of deer viruses 1 and 2 into sheep at 40, 60, and 80 days gestation resulted in the mummification of two fetuses and one abortion, but the lambs born normally had no defects, were not viremic, and had no serum-neutralizing antibodies to any of the viruses (PARSONSON and LUEDKE, unpublished).

HARE et al. (1988) infected donor sheep with BTV-11 (Texas station strain) bred to BTV-infected and non-infected rams. A total of 49 embryos were collected from 18 BTV-infected ewes and transferred to 27 BTV-seronegative recipient ewes. Eleven pregnancies and 12 lambs resulted. None of the lambs or recipients had serum antibodies nor was BTV isolated from any animal at slaughter 30 days after parturition.

6 Effect of Bluetongue Virus on Reproduction in Cattle

The transmission of BTV vertically via the infected pregnant dam to the fetus was postulated by LUEDKE et al. (1970). Inoculation of fetal calves at the 5th month of gestation resulted in failure to isolate BTV at birth although the calves were normal and most had neutralizing antibodies in their sera (JOCHIM et al. 1974).

Using *Culicoides variipennis* infected with BTV (LUEDKE et al. 1977) demonstrated abortions and congenital anomalies in calves from cattle exposed at 60 and 120 days pregnant. Two abortions occurred and one fetus was dead at parturition. The remaining seven live calves had gross anomalies and BTV was isolated from four. One of the male calves was later found to excrete BTV in semen and to have a persistent viremia from birth through 11 years (LUEDKE et al. 1982). At 5 years of age the bull was used in a natural breeding experiment and was mated with 14 cows (LUEDKE and WALTON 1980). Twelve live calves were born and one calf was dead at birth. The 12 live calves had congenital defects compatible with life and all had a sporadic persistent viremia without producing precipitating or neutralizing antibodies to BTV. LUEDKE (1985) described their subsequent history and suggested that survival of these calves under field conditions would have been unlikely.

The same bull (B28A) was used to inseminate four cows at estrus. Additional groups of four cows each were inoculated subcutaneously with BTV isolated from the bull (BTV-11) at various stages of pregnancy, or inoculated directly into the uterus at estrus during insemination. None of the four cows mated to the bull nor inoculated into the uterus with BTV-11 became infected. Seven of the eight remaining animals had viremia, produced antibodies to BTV and had fetuses ranging in age from 70 to 222 days which appeared grossly normal. BTV was not isolated from the fetuses and those capable of mounting a humoral immune response had no antibodies in their serums. It was concluded that there was no evidence for congenital BTV-11 infection in this study (PARSONSON and LUEDKE, unpublished).

Studies have been carried out in cattle in which BTV has been used to infect cows at insemination (GROOCOCK et al. 1983; BOWEN and HOWARD 1984; PARSONSON et al. 1987a). In the studies animals became infected as evidenced by viremia and the production of antibodies to the respective BTV used. No clinical disease was observed, virus was not isolated from tissues of cattle slaughtered, and no evidence of fetal infection was detected in the calves that resulted from the insemination.

PARSONSON et al. (1987b) inoculated BTV-20 into three groups of cows pregnant at 84 to 95, 100 to 160, and 170 to 180 days. No clinical signs were observed, either viremia and/or antibodies were produced in 11 of the 12 animals. Cows, calves, and fetuses were necropsied following either parturition or slaughter between 200 or 270 days of pregnancy. No virus isolations were made from a wide range of tissues and no immunoglobulins or serum-neutralizing antibodies to bluetongue virus were detected in the serums of calves or fetuses. It was concluded that there was no evidence of transplacental transfer of BTV-20. BOWEN et al. (1982), exposed preimplantation embryos from mice and cattle to BTV-11 or BTV-17. The zona pellucida (ZP) of both murine and bovine embryos provided effective protection from virus present in culture fluid. However, ZP-free embryos became infected and the BTV infection was rapidly embryocidal. SINGH et al. (1982) carried out similar studies with bovine embryos and found the ZP afforded protection to BTV.

In a later study (BOWEN et al. 1983a) collected embryos from donor cattle using nonsurgical methods at the peak of BTV infection and transferred the embryos to 39 recipients of which 21 became pregnant. No transmission of BTV from the infected donors to uninfected recipients occurred. In another study THOMAS et al. (1983) transferred 28 ZP-intact embryos from cows infected with either BTV-17 or BTV-18 to noninfected recipient cows and 14 became pregnant. No recipient developed a viremia or antibodies to BTV and no virus or antibody was detected in the calves at parturition.

BTV has been detected in the semen of bulls infected with the virus, (LUEDKE et al. 1975; BRECKON et al. 1980; PARSONSON et al. 1981; BOWEN et al. 1983b). However, while LUEDKE et al. (1975) and BRECKON et al. (1980) were able to isolate virus from the semen of persistently infected bulls, PARSONSON et al. (1981), BOWEN et al. (1983b; 1985), and HOWARD et al. (1985) isolated BTV from semen only during the time that the bull had a concurrent viremia (approximately 14 to 28 days). BTV was not isolated from the semen of the majority of bulls in these experimental groups.

As part of the management of some artificial breeding (AB) centers, regular serological tests and virus isolation attempts to monitor for virus infections are carried out (PHILLIPS et al. 1986; MONKE et al. 1986). Bulls with and without serum antibodies to BTV were present on the centers. Bulls seropositive for BTV infection were monitored for excretion of BTV over 3 and 4.5 years (PHILLIPS et al. 1986; MONKE et al. 1986). In both studies all attempts to isolate BTV from semen samples from seropositive and seronegative bulls were unsuccessful. It was concluded from these results that positive serology for antibodies to BTV is poorly correlated with viremia and would indicate that persistent BTV viremias are unlikely. One bull that was reported to be persistently infected with BTV (-13 and -11) and to shed BTV in his semen, was the subject of a retrospective study (LUEDKE et al. 1982). The serological and virological responses of this 11-year-old bull were detailed in an effort to eliminate some of the confusion surrounding his past history. The bull had serum-neutralizing antibodies to BTV from 6 months of age, and would have been detected as seropositive in an AB center.

Embryo transfer has recently been used as a means of controlling BTV infection in cattle resulting from insemination of infected semen. THOMAS et al. (1985) inseminated four heifers with semen containing BTV-17, and 20 embryos were collected and transferred to 16 recipients. None of the recipients or calves resulting developed BTV infection or antibody, although two of the donor cows had viremia at the time of embryo collection.

6.1 Congenital Bluetongue Virus Infections in Cattle

In a series of experiments (LUEDKE et al. 1977; LUEDKE and WALTON 1980) suggested that BTV was abortigenic and teratogenic in cattle. There has been little documentary evidence from Africa, Australia, or the USA where BTV are endemic in field situations to substantiate these claims. Many experimental studies have used direct inoculation of the bovine fetus in utero to demonstrate teratogenicity of BTV (BARNARD and PIENAAR 1976; MACLACHLAN and OSBURN

1983; MacLachlan et al. 1985; Thomas et al. 1986). All the studies found that BTV infections in the early bovine fetus resulted in severe central nervous system destruction. Any surviving calves would have poor viability and would not be of significance as reservoirs of BTV. Congenital infection did not lead to specific immunologic tolerance or to postnatal persistence of virus.

7 Bluetongue Virus Infection in Other Species

7.1 Goats

BTV infection of goats is usually inapparent or mild (Erasmus 1975; Barzilai and Tadmor 1971; Luedke and Anakwenze 1972). Clinical bluetongue disease has been reported in goats in India (Sapre 1964). The disease in goats has warranted very little comment in the literature, although goats have been known to be susceptible for over 80 years (Spreull 1905).

7.2 Wildlife Species in the USA

Hoff and Trainer (1978) have reviewed the effects of BTV and EHD virus on wildlife in the USA. They concluded that EHD appeared to be primarily a disease of white-tailed deer, although virus-specific antibodies have been detected in cattle and sheep (Metcalf et al. 1981).

BTV appears to be transmitted from domestic livestock to wildlife species and acute to peracute fatal hemorrhagic diseases often occur in white-tailed deer. Karsted and Trainer (1967) and Vosdingh et al. (1968) found that the clinical disease and lesions in white-tailed deer to BTV and EHD virus were similar and resulted in high mortality.

Both BTV and EHD virus have been isolated from a number of free-ranging wild ruminants in the USA. Thorne et al. (1988), described two epizootics in pronghorn antelope (*Antilocapea americana*) in Wyoming during which it was estimated 3200 animals died in 1976 and a further 300 in 1984. Mule deer (*Odocoileus hemionus*) deaths also occurred at the same time from what was presumed to be bluetongue disease.

7.3 Wildlife Species in Africa

Neitz (1933) first demonstrated BTV infection in the blesbuck (*Damaliscus albifrons*); subsequently many wild ruminants in Africa have been shown to be natural hosts for BTV (Erasmus 1980). Large populations of wild ruminants provide the basis for the maintenance cycle for persistence of BTV in Kenya

(DAVIES 1980). Examination of sera from a wide range of wild ruminants showed a high prevalence of antibodies to BTV.

8 Epizootic Hemorrhagic Disease (EHD) of Deer Virus Infection in Cattle

The importance of EHD virus infection of cattle and wild ruminant populations has only been assessed in Japan and the USA where epizootics in cattle (Japan) and in deer (USA) have occurred. The relationship of EHD virus and BTV infections in ruminants in endemic areas of the world is yet to be investigated to identify the interrelationship and affects of common infections by both virus types in ruminants.

8.1 EHD Virus in Australia

There are at least five serotypes of EHD virus in Australia known to infect cattle, buffalo, and deer without causing clinical disease (ST. GEORGE 1985).

8.2 EHD Virus in Japan

Descriptions of Ibaraki disease in Japanese cattle (INABA 1975) were originally compared with descriptions of bluetongue in cattle in Portugal and Spain, however, Ibaraki virus did not cause disease in sheep. Ibaraki virus is now known to be EHD virus serotype 2 and to be serologically similar to EHDV-2 (Alberta strain) and the Australian isolate CSIRO 439 (CAMPBELL and ST. GEORGE 1986).

Serological surveys of Japanese cattle have identified the presence of antibodies to several BTV serotypes including BTV-1, BTV-2, and BTV-20 (MIURA et al. 1982a, b) without evidence of clinical disease to either EHD virus or BTV.

8.3 EHD Virus in the United States

Following a serological survey of slaughter cattle in the USA (METCALF et al. 1981), it was found that prevalence of serum antibodies to BTV was low in northern states and high in southwestern states, ranging from 0% to 79% of samples tested. The bluetongue antibody-positive sera were further analyzed to determine the distribution of serotypes. Although not all the sera were tested for neutralizing antibody to EHD virus, there were strong indications that EHD virus was present in every region and was a relatively common infection of cattle, (CARLSON 1983).

The presence of mixed orbivirus infections in cattle and some wild ruminants in endemic areas of the world may ensure a constant immunological stimulus for their protection against disease outbreaks and at the same time provide a reservoir for periodic epizootics of both BTV and EHD virus.

9 Conclusions

Many puzzling aspects of the pathogenesis of BTV and related orbiviruses in ruminants remain to be solved. However, the great increase in information on the BTV genome and the virion structure and its proteins, as well as the exquisite studies on viral replication within cells, have provided new tools for further studies on the vector-host-parasite relationships of this virus genus and its infections.

The pathology and pathogenesis of BTV in sheep require further study to enable better assessment of the role of virulence factors and the development of immunizing methods in order to protect sheep flocks in endemic areas and against potential epizootics outside endemic areas.

The role of BTV in reproduction in both sheep and cattle has been an area of confusion and of international concern for trade in domestic ruminant genetic material. A gradual accumulation of scientific data will enable authorities to develop rational approaches to disease control within their own countries and allow relaxation of restrictive international requirements. Advances in the development of superior animal types, better methods for screening and handling gametes, and diagnostic systems more acceptable internationally will result from such studies.

References

Anderson CK, Jensen R (1969) Pathologic changes in placentas of ewes inoculated with bluetongue virus. Am J Vet Res 30: 987–999

Anderson GA, Stott JL, Gershwin LJ, Osburn BI (1985) Subclinical and clinical bluetongue disease in cattle: clinical, pathological and pathogenic considerations. In: Barber TL, Jochim MM (eds) Bluetongue and related orbiviruses. Liss, New York, pp 103–107

Appleton JA, Letchworth GJ (1983) Monoclonal antibody analysis of serotype-restricted and unrestricted bluetongue viral antigenic determinants. Virology 124: 286–299

Barnard BJH, Pienaar JG (1976) Bluetongue virus as a cause of hydranencephaly in cattle. Onderstepoort J Vet Res 43: 155–158

Barzilai E, Shimshony A (1985) Bluetongue: virological and epidemiological observations in Israel. In: Barber TL, Jochim MM (eds) Bluetongue and related orbiviruses. Liss, New York, pp 545–553

Barzilai E, Tadmor A (1971) Multiplication of bluetongue virus in goats following experimental infection. Refuah Vet 28: 11–20

Bekker JG, De Kock GW, Quinlan JB (1934) The occurrence and identification of bluetongue in cattle—the so-called pseudo-foot and mouth disease in South Africa. Onderstepoort J Vet Sci Anim Indust 2: 393–507
Bowen RA, Howard TH (1984) Transmission of bluetongue virus by intrauterine inoculation or insemination of virus-containing bovine semen. Am J Vet Res 45: 1386–1388
Bowen RA, Howard TH, Pickett BW (1982) Interaction of bluetongue virus with preimplantation embryos from mice and cattle. Am J Vet Res 43: 1907–1911
Bowen RA, Howard TH, Elsden RP, Seidel GE (1983a) Embryo transfer from cattle infected with bluetongue virus. Am J Vet Res 44: 1625–1628
Bowen RA, Howard TH, Entwistle KW, Pickett BW (1983b) Seminal shedding of bluetongue virus in experimentally infected mature bulls. Am J Vet Res 44: 2268–2270
Bowen RA, Howard TH, Pickett BW (1985) Seminal shedding of bluetongue virus in experimentally infected bulls. In: Barber TL, Jochim MM (eds) Bluetongue and related orbiviruses. Liss, New York, pp 91–96
Bowne JG, Luedke AJ, Jochim MM, Metcalf HE (1968) Bluetongue disease in cattle. J Am Vet Med Assoc 153: 662–668
Breckon RD, Luedke AJ, Walton TE (1980) Bluetongue virus in bovine semen: viral isolation. Am J Vet Res 41: 439–442
Bwangomi O (1978) Pathology of ovine fetus infection with bluetongue virus. Bull Anim Health Prod Afr 26: 79–97
Campbell CH, St George TD (1968) A preliminary report of a comparison of epizootic hemorrhagic disease viruses from Australia with others from North America, Japan and Nigeria. Aust Vet J 63: 233
Carlson JH (1983) Differential serotyping of bluetongue virus—positive serums obtained in a national survey. In: Compans RW, Bishop DHL (eds) Double stranded RNA viruses. Elsevier, New York, pp 361–366
Cordy DR, Schultz G (1961) Congenital subcortical encephalopathies in lambs. J Neuropathol Exp Neurol 20: 554–562
Cowly JA, Gorman BM (1987) Genetic reassortants for identification of the genome segment coding for the bluetongue virus hemagglutinin. J Virol 61: 2304–2306
Davies FG (1978) Bluetongue studies with sentinel cattle in Kenya. J Hyg 80: 197–204
Davies FG (1980) Bluetongue in Kenya. Bull Off Int Epiz 92: 469–481
De Kock G, Du Toit R, Neitz WO (1937) Observations on blue tongue in cattle and sheep. Onderstepoort J Vet Sci Anim Ind 8: 129–180
Della-Porta AJ, Sellers RF, Herniman KAJ, Littlejohns IR, Cybinski DH, St. George TD, McPhee DA, Snowdon WA, Campbell J, Cargill C, Corbould A, Chung YS, Smith VW (1983) Serological studies of Australian and Papua New Guinean cattle and Australian sheep for the presence of antibodies against bluetongue group viruses. Vet Microbiol 8: 147–162
Dixon RW (1909) Catarrhal fever of sheep. Agric J Cape Good Hope 34: 486–491
Eaton BT, Crameri GF (1989) The site of bluetongue virus attachment to glycophorins from a number of animal erythrocytes. J Gen Virol (in press)
Enright FM, Osburn BI (1980) Ontogeny of host responses in ovine fetuses infected with bluetongue virus. Am J Vet Res 41: 224–229
Erasmus BJ (1975) Bluetongue in sheep and goats. Aust Vet J 51: 165–170
Erasmus BJ (1980) The epidemiology and control of bluetongue in South Africa. Bull Off Int Epiz 92: 461–467
Flanagan M, Wilson AJ, Trueman KF, Shepherd MA (1982) Bluetongue virus serotypes 20 infection in pregnant Merino sheep. Aust Vet J 59: 18–20
Gambles RM (1949) Bluetongue of sheep in Cyprus. J Comp Pathol 59: 176–190
Gibbs EPJ, Lawman MJP, Herniman KAJ (1979) Preliminary observations on transplacental infection of bluetongue virus in sheep—a possible overwintering mechanism. Res Vet Sci 27: 118–120
Gibbs EJP, Greiner EC (1985) Serological observations on the epidemiology of bluetongue virus infections in the Caribbean and Florida. In: Barber TL, Jochim MM (eds) Bluetongue and related orbiviruses. Liss, New York, pp 563–570
Graf H (1933) Chemical blood studies III. Comparative studies on "laked" and "unlaked" blood filtrates of sheep in health and during "heartwater" (*Rickettsia ruminantium* infection) and bluetongue (catarrhal fever). Onderstepoort J Vet Sci Anim Ind 1: 285–334
Groocock CM, Parsonson IM, Campbell CH (1982) Bluetongue virus serotypes 20 and 17 infections in sheep: comparison of clinical and serological responses. Vet Microbiol 7: 189–196

Groocock CM, Parsonson IM, Campbell CH (1983) Bluetongue virus serotype 20 infections in cattle of breeding age. Am J Vet Res 44: 1765–1768
Gumm ID, Taylor WP, Roach CJ, Alexander FCM, Greiner EC, Gibbs EPJ (1984) Serological survey of ruminants in some Caribbean and South American countries for type-specific antibody to bluetongue and epizootic hemorrhagic disease viruses. Vet Rec 114: 635–638
Hafez SM, Taylor WP (1985) Serotypes of bluetongue virus present in Saudi Arabia. In: Barber TL, Jochim MM (eds) Bluetongue and related orbiviruses. Liss, New York, pp 531–537
Hardy WT, Price DA (1952) Soremuzzle of sheep. J Am Vet Med Assoc 120: 23–25
Hare WCD, Luedke AJ, Thomas FC, Bowen RA, Singh EL, Eaglesome MD, Randall GCB, Bielanski A (1988) Nontransmission of bluetongue virus by embryos from bluetongue virus-infected sheep. Am J Vet Res 49: 468–472
Heidner HW, MacLachlan NJ, Fuller FJ, Richards RG, Whetter LE (1988) Bluetongue virus genome remains stable throughout prolonged infection of cattle. J Gen Virol 69: 2629–2636
Herniman KAJ, Gumm ID, Owen L, Taylor WP, Sellers RF (1980) Distribution of bluetongue viruses and antibodies in some countries of the Eastern Hemisphere. Bull Off Int Epiz 92: 581–586
Hoff GL, Trainer DO (1978) Bluetongue and epizootic hemorrhagic disease viruses: their relationship to wildlife species. Adv Vet Sci Comp Med 22: 111–132
Homan EJ, Lorbacher de RH, Donato A, Taylor W, Yuill TM (1985) Bluetongue virus infection in Costa Rican and Colombian cattle. In: Barber TL, Jochim MM (eds) Bluetongue and related orbiviruses. Liss, New York pp 559–561
Hourrigan JL, Klingsporn AL (1975) Bluetongue: the disease in cattle. Aust Vet J 51: 170–174
Howard TH, Bowen RA, Pickett BW (1985) Isolation of bluetongue virus from bull semen. In: Barber TL, Jochim MM (eds) Bluetongue and related orbiviruses. Liss, New York, pp 127–134
Howell PG (1960) A preliminary antigenic classification of strains of bluetongue virus. Onderstepoort J Vet Res 28: 357–364
Howell PG (1963) Emerging diseases of animals. FAO Agricultural Studies, no. 61, Rome, pp 109–154
Howell PG (1970) The antigenic classification and distribution of naturally occurring strains of bluetongue virus. J S Afr Vet Med Assoc 41: 215–223
Howell PG, Verwoerd DW, Oellermann RA (1967) Plaque formation by bluetongue virus. Onderstepoort J Vet Res 34: 317–332
Huismans H, Erasmus BJ (1981) Identification of the serotype-specific and group-specific antigens of bluetongue virus. Onderstepoort J Vet Res 48: 51–58
Huismans H, van der Walt NT, Cloete M, Erasmus BJ (1983) The biochemical and immunological characterisation of bluetongue virus outer capsid polypeptides In: Compans RW, Bishop DHL (eds) Double-stranded RNA viruses. Elsevier, New York, pp 165–172
Huismans H van der Walt NT, Cloete M, Erasmus BJ (1987) Isolation of a capsid protein of bluetongue virus that induces a protective immune response in sheep. Virology 157: 172–179
Inaba Y (1975) Ibaraki disease and its relationship to bluetongue. Aust Vet J 51: 178–185
Jeggo MH, Gumm ID, Taylor WP (1983) Clinical and serological response of sheep to serial challenge with different bluetongue virus types. Res Vet Sci 34: 205–211
Jeggo MJ, Corteyn AH, Taylor WP, Davidson WH, Gorman BM (1986) Virulence of bluetongue virus for British sheep. Res Vet Sci 42: 24–28
Jochim MM, Luedke AJ, Chow TL (1974) Bluetongue in cattle: immunogenic and clinical responses in calves inoculated in utero and after birth. Am J Vet Res 35: 517–522
Karstad L, Trainer DO (1967) Histopathology of experimental bluetongue disease of white-tailed deer. Can Vet J 8: 247–254
Komarov A, Goldsmit L (1951) A disease, similar to bluetongue in cattle and sheep in Israel. Refuah Vet 8: 69–74
Lawman MJP (1979) Observations on the pathogenesis of bluetongue virus infections in sheep. PhD thesis, University of Surrey, England
Letchworth GJ, Appleton JA (1983) Heterogeneity of neutralization-related epitopes within a bluetongue virus serotype. Virology 124: 300–307
Lopez AC, Sanchez Botija C (1958) L'epizootie de fievre catarrhale ovine en Espagne (Blue tongue). Bull Off Int Epiz 50: 65–93
Luedke AJ (1969) Bluetongue in sheep: viral assay and viremia. Am J Vet Res 30: 499–509
Luedke AJ (1970) Distribution of virus in blood components during viremia of bluetongue. Proc Annu Meet US Anim Health Assoc 74: 9–21
Luedke AJ (1985) Effect of bluetongue virus on reproduction in sheep and cattle. In: Barber TL, Jochim MM (eds) Bluetongue and related orbiviruses. Liss, New York, pp 71–78
Luedke AJ, Anakwenze EI (1972) Bluetongue virus in goats. Am J Vet Res 33: 1739–1745

Luedke AJ, Jochim MM (1968) Bluetongue virus in sheep. Intensification of the clinical response by previous oral administration. Cornell Vet 58: 48–58
Luedke AJ, Walton TE (1980) Effect of natural breeding of heifers to a bluetongue virus carrier bull. Proc. 11th Int Congress on Diseases of Cattle, Tel-Aviv, Israel, pp 478–491
Luedke AJ, Bowne JG, Jochim MM, Doyle C (1964) Clinical and pathologic features of bluetongue in sheep. Am J Vet Res 25: 963–970
Luedke AJ, Jochim MM, Bowne JG, Jones RH (1970) Observations on latent bluetongue virus infection in cattle. J Am Vet Med Assoc 156: 1871–1879
Luedke AJ, Walton TE, Jones RH (1975) Detection of bluetongue virus in bovine semen. Proc 20th World Vet Congress, Thessaloniki, Greece, vol 3, pp 2039–2042
Luedke AJ, Jochim MM, Jones RH (1977) Bluetongue in cattle: effects of *Culicoides variipennis* transmitted bluetongue virus on pregnant heifers and their calves. Am J Vet Res 38: 1687–1695
Luedke AJ, Jochim MM, Barber TL (1982) Serologic and virologic responses of a Hereford bull persistently infected with bluetongue virus for eleven years. Am Assoc Vet Lab Diagn Annu Proc 25: 115–133
MacLachlan NJ, Fuller FJ (1986) Genetic stability in calves of a single strain of bluetongue virus. Am J Vet Res 47: 762–764
MacLachlan NJ, Osburn BI (1983) Bluetongue virus-induced hydranencephaly in cattle. Vet Pathol 20: 563–573
MacLachlan NJ, Osburn BI, Ghalib HW, Stott JL (1985) Bluetongue virus-induced encephalopathy in fetal cattle. Vet Pathol 22: 415–417
MacLachlan NJ, Heidner HW, Fuller FJ (1987) Humoral immune response of calves to bluetongue virus. Am J Vet Res 48: 1031–1035
Mahrt CR, Osburn BI (1986) Experimental bluetongue virus infection of sheep; effect of vaccination: pathologic, immunofluorescent, and ultrastructural studies. Am J Vet Res 47: 1198–1203
Manso-Ribeiro J, Rosa-Azevedo JA, Noronha FO, Braco-Forte MC, Grave-Pereira C, Vasco-Fernandes M (1957) Fievre catarrhale du mouton (Blue-tongue). Bull Off Int Epiz 48: 350–367
Mason JH, Neitz WO (1940) The susceptibility of cattle to the virus of bluetongue. Onderstepoort J Vet Sci Anim Ind 15: 149–157
McKercher DG, McGowan B, Howarth JA, Saito JK (1953) A preliminary report on the isolation and identification of the bluetongue virus from sheep in California. J Am Vet Med Assoc 122: 300–301
McPhee DA, Parsonson IM, Della-Porta AJ (1982) Comparative studies on the growth of Australian bluetongue virus serotypes in continuous cell lines and embryonated chicken eggs. Vet Microbiol 7: 401–410
Metcalf HE, Pearson JE, Klingsporn AL (1981) Bluetongue in cattle: a serologic survey of slaughter cattle in the United States. Am J Vet Res 42: 1057–1061
Miura Y, Inaba Y, Tsuda T, Tokuhisa S, Sato K, Akashi H (1982a) Seroepizootiological survey on bluetongue virus infection in cattle in Japan. Natl Inst Anim Health Q 22: 154–158
Miura Y, Inaba Y, Tsuda T, Tokuhisa S, Sato K, Akashi H, Matumoto M (1982b) A survey of antibodies to arthropod-borne viruses in Indonesian cattle. Jpn J Vet Sci 44: 857–863
Monke DR, Hueston WD, Call JW (1986) Veterinary management of an artificial insemination center containing bluetongue seropositive bulls. Proc Annu Meet US Anim Health Assoc 90: 139–143
Morrill JC, McConnell S (1985) An electron microscopic study of blood cells from calves experimentally infected with bluetongue virus. In: Barber TL, Jochim MM (eds) Bluetongue and related orbiviruses. Liss, New York, pp 279–287
Moulton JE (1961) Pathology of bluetongue of sheep in California. J Am Vet Med Assoc 138: 493–498
Muller MJ, Standfast HA, St. George TD, Cybinski DH (1982) *Culicoides brevitarsis* (Diptera: Ceratopogonidae) as a vector of arboviruses in Australia. Proc 3rd Arbovirus Symposium, Brisbane, pp 43–49
Murray MD (1987) Akabane epizootics in New South Wales: evidence for long-distance dispersal of the biting midge *Culicoides brevitarsis*. Aust Vet J 64: 305–308
Neitz WO (1933) The blesbuck (*Damaliscus albifrons*) as a carrier of heartwater and bluetongue. J S Afr Vet Med Assoc 4: 24–26
Neitz WO, Riemerschmid G (1944) The influence of solar radiation on the course of bluetongue. Ondestepoort J Vet Sci Anim Ind 20: 29–56
Osburn BI, Silverstein AM, Prendergast RA, Johnson RT, Parshall CJ (1971a) Experimental viral-induced congenital encephalopathies. I. Pathology of hydranencephaly and porencephaly caused by bluetongue vaccine virus. Lab Invest 25: 197–205

Osburn BI, Johnson RT, Silverstein AM, Prendergast RA, Jochim MM, Levy SE (1971b) Experimental viral-induced congenital encephalopathies. II. The pathogenesis of bluetongue vaccine virus infection in fetal lambs. Lab Invest 25: 206–210
Owen NC, Du Toit RM, Howell PG (1965) Bluetongue in cattle: typing of viruses isolated from cattle exposed to natural infections. Onderstepoort J Vet Res 32: 3–6
Ozawa Y (1985) Bluetongue and related orbiviruses: overview of the world situation. In: Barber TL, Jochim MM (eds) Bluetongue and related orbiviruses. Liss, New York, pp 13–20
Parish SM, Evermann JF, Olcott B, Gay C (1982) A bluetongue epizootic in northwestern United States. J Am Vet Med Assoc 181: 589–591
Parsonson IM, Snowdon WA (1985) Bluetongue, epizootic hemorrhagic disease of deer and related viruses: current situation in Australia. In: Barber TL, Jochim MM (eds) Bluetongue and related orbiviruses. Liss, New York, pp 27–35
Parsonson IM, Della-Porta AJ, McPhee DA, Cybinski DH, Squire KRE, Standfast HA, Uren MF (1981) Isolation of bluetongue virus serotype 20 from the semen of an experimentally-infected bull. Aust Vet J 57: 252
Parsonson IM, Della-Porta AJ, McPhee DA, Cybinski DH, Squire KRE, Uren MF (1987a) Experimental infection of bulls and cows with bluetongue virus serotype 20. Aust Vet J 64: 10–13
Parsonson IM, Della-Porta AJ, McPhee DA, Cybinski DH, Squire KRE, Uren MF (1987b) Bluetongue virus serotype 20: experimental infection of pregnant heifers. Aust Vet J 64: 14–17
Phillips RM, Carnahan DL, Rademacher DJ (1986) Virus isolation from semen of bulls serologically positive for bluetongue virus. Am J Vet Res 47: 84–85
Pini A (1976) A study on the pathogenesis of bluetongue: replication of the virus in the organs of infected sheep. Onderstepoort J Vet Res 43: 159–164
Pini A, Coackley W, Ohder H (1966) Concentration of bluetongue virus in experimentally infected sheep and virus identification by immune fluorescence technique. Arch Gesamt Virusforsch 18: 385–390
Polydorou K (1985) Bluetongue in Cyprus. In: Barber TL, Jochim MM (eds) Bluetongue and related orbiviruses. Liss, New York, pp 539–544
Radostits OM, Littlejohns IR (1988) New concepts in the pathogenesis, diagnosis and control of diseases caused by the bovine viral diarrhoea virus. Can Vet J 29: 513–528
Richards RG, MacLachlan NJ, Heidner HW, Fuller FJ (1988) Comparison of virologic and serologic responses of lambs and calves infected with bluetongue virus serotype 10. Vet Microbiol 18: 233–242
Richards WPC, Cordy DR (1967) Bluetongue virus infection: pathologic responses of nervous systems in sheep and mice. Science 156: 530–531
Sapre SN (1964) An outbreak of "bluetongue" in goats and sheep in Maharashtra State, India. Vet Rev 15: 69–71
Sawyer MM, Osburn BI (1977) Isolation of bluetongue vaccine virus from infected sheep by direct inoculation onto tissue culture. Cornell Vet 67: 65–71
Sellers RF (1975) Bluetongue in Cyprus. Aust Vet J 51: 198–203
Shimshony A, Goldsmit L, Barzilai E (1980) Bluetongue in Israel. Bull Off Int Epiz 92: 525–534
Shimshony A, Barzilai E, Savir D, Davidson M (1988) Epidemiology and control of bluetongue disease in Israel. Rev Sci Tech Off Int Epiz 7: 311–329
Shultz D, DeLay PD (1955) Losses in newborn lambs associated with bluetongue vaccination of pregnant ewes. J Am Vet Med Assoc 127: 224–226
Silva FS (1956) Linguá azul ou febre catarral dos ovinos (bluetongue). Rev Ciencia Vet (Lisboa) 51: 191–231
Singh EL, Eaglesome MD, Thomas FC, Papp-Vid G, Hare WCD (1982) Embryo transfer as a means of controlling the transmission of viral infections. I. The in vitro exposure of preimplantation bovine embryos to Akabane, bluetongue and bovine viral diarrhea viruses. Theriogenology 17: 437–444
Spreull J (1905) Malarial catarrhal fever (bluetongue) of sheep in South Africa. J Comp Pathol 18: 321–337
Stair EL (1968) The pathogenesis of bluetongue in sheep: a study by immunofluorescence and histopathology. PhD thesis, Texas A & M University, USA
St. George TD, McCaughan CI (1979) The transmission of the CSIRO19 strain of bluetongue virus type 20 to sheep and cattle. Aust Vet J 55: 198–199
St. George TD (1985) Epidemiology of bluetongue in Australia: the vertebrate hosts. In: Barber TL, Jochim MM (eds) Bluetongue and related orbiviruses. Liss, New York, pp 519–525
Stott JL, Anderson GA, Jochim MM, Barber TL, Osburn BI (1982) Clinical expression of bluetongue disease in cattle. Proc Annu Meet US Anim Health Assoc 86: 126–131

Tamayo R, Schoebitz R, Alonso O, Wenzel J (1985) First report of bluetongue antibody in Chile. In: Barber TL, Jochim MM (eds) Bluetongue and related orbiviruses. Liss, New York, pp 555–558

Taylor WP, Sellers RF, Gumm ID, Herniman KAJ, Owen L (1985) Bluetongue epidemiology in the Middle East. In: Barber TL, Jochim MM (eds) Bluetongue and related orbiviruses. Liss, New York, pp 527–530

Theiler A (1906) Bluetongue in sheep. Annual report, Director of Agriculture Transvaal (1904–1905), pp 115–121

Thomas AD, Neitz WO (1947) Further observations on the pathology of bluetongue in sheep. Onderstepoort J Vet Sci Anim Indust 22: 27–40

Thomas FC, Skinner DJ, Samagh BS (1982) Evidence for bluetongue virus in Canada: 1976–1979. Can J Comp Med 46: 350–353

Thomas FC, Singh EL, Hare WCD (1983) Embryo transfer as a means of controlling viral infections. III. Non transmission of bluetongue virus from viremic cattle. Theriogenology 19: 425–431

Thomas FC, Singh EL, Hare WCD (1985) Embryo transfer as a means of controlling viral infections. VI. Bluetongue virus-free calves from infectious semen. Theriogenology 24: 345–350

Thomas FC, Randall GCB, Myers DJ (1986) Attempts to establish congenital bluetongue virus infections in calves. Can J Vet Res 50: 280–281

Thorne ET, Williams ES, Spraker TR, Helms W, Segerstrom T (1988) Bluetongue in free-ranging pronghorn antelope (*Antilocapra americana*) in Wyoming: 1976 and 1984. J Wildl Dis 24: 113–119

Uren MF, St. George TD (1985) The clinical susceptibility of sheep to four Australian serotypes of bluetongue virus. Aust Vet J 62: 175–176

Uren MF, Squire KRE (1982) The clinico-pathological effect of bluetongue virus serotype 20 in sheep. Aust Vet J 58: 11–15

Vosdingh RA, Trainer DO, Easterday BC (1968) Experimental bluetongue disease in white-tailed deer. Can J Comp Med Vet Sci 32: 382–387

Wirth WW, Dyce AL (1985) The current taxonomic status of the *Culicoides* vectors of bluetongue viruses. In: Barber TL, Jochim MM (eds) Bluetongue and related orbiviruses. Liss, New York, pp 151–164

Young S, Cordy DR (1964) An ovine fetal encephalopathy caused by bluetongue vaccine virus. J Neuropathol Exp Neurol 23: 635–659

The Replication of Bluetongue Virus in *Culicoides* Vectors

P. S. MELLOR

1 Introduction 143
2 Vector Species of *Culicoides* 144
2.1 Africa and the Middle East 145
2.2 Asia 145
2.3 Australia 146
2.4 The Americas 147
2.5 Europe 147
3 Infection of *Culicoides* with Bluetongue Viruses 148
3.1 Parenteral Inoculation 148
3.2 Oral Infection and Replication of BTV in Susceptible *Culicoides* 150
3.3 Persistence of Infection 152
3.4 Oral Susceptibility of *C. Variipennis* to Different Bluetongue Serotypes 152
4 Barriers to the Infection of Arthropods with Viruses 153
4.1 Barriers to the Infection of *Culicoides* with Bluetongue Virus 155
5 Other Factors Involved in Bluetongue Virus Infection of *Culicoides* 156
6 Summary 157
References 158

1 Introduction

Bluetongue virus (BTV) has long been known to be transmitted biologically by certain species of biting midge belonging to the genus *Culicoides* (Latreille). DU TOIT (1944) in South Africa was the first to implicate a *Culicoides* species in the transmission of this virus when he showed that *C. imicola* (*pallidipennis*) was able to transmit bluetongue from infected to susceptible sheep. Since that time numerous authors have confirmed his observations (WALKER and DAVIES 1971; BRAVERMAN and GALUN 1973; BRAVERMAN et al. 1981, 1985; MELLOR et al. 1984a). However, although there are well over 1000 species of *Culicoides* in the world (BOORMAN 1988), only 17 have been connected with BTV and to date only six, *C. variipennis*, *C. imicola*, *C. fulvus*, *C. actoni*, *C. wadai* and *C. nubeculosus*, have been proven to transmit the virus (Table 1). This number may soon be increased to eight since on the basis of epidemiological evidence and virus isolations it is

AFRC Institute for Animal Health, Pirbright Laboratory, Ash Road, Pirbright, Nr. Woking, Surrey GU24 0NF, UK

Current Topics in Microbiology and Immunology, Vol. 162

Table 1. Field and laboratory BTV infections of *Culicoides*

Subgenus	Species	Virus isolation	Laboratory infection	Transmission
Avaritia	*C. actoni*	−	+	+
	C. brevipalpis	−	+	−
	C. brevitarsis	+	+	−
	C. fulvus	+	+	+
	C. imicola	+	+	+
	C. obsoletus	+	−	−
	C. tororoensis	+	−	−
	C. wadai	−	+	+
Culicoides	*C. peregrinus*	−	±	−
Diphaeomia	*C. debilipalpis*	−	+	−
Hoffmania	*C. insignis*	+	−	−
	C. milnei	+	−	−
	C. venustus	−	+	−
Monoculicoides	*C. variipennis*	+	+	+
	C. nubeculosus	−	+	+
Oecacta	*C. oxytoma*	−	±	−
Pulicaris	*C. impunctatus*	−	±	−

likely that *C. insignis* and *C. brevitarsis* will also prove to be competent BTV vectors (GREINER et al. 1985; STANDFAST et al. 1985). Although some of the remaining species of *Culicoides* may eventually be shown to be fully competent BTV vectors most species will be refractory to infection. Why this should be is not yet entirely clear, but it is known that the mechanism or mechanisms controlling the oral infection of *Culicoides* with BTV operate chiefly at the level of the mid-gut wall (JENNINGS and MELLOR 1988), a single layer of cells of epithelial origin supported by a basement lamina (HARDY et al. 1983; MEGAHED 1956).

The aims of this chapter are: (a) to discuss those species of *Culicoides* which are either confirmed or suspected vectors of BTV in the various parts of the world where the virus occurs; (b) to review the studies which have been undertaken into bluetongue virogenesis in vector species of *Culicoides*, and (c) to discuss some of the factors which may be involved in controlling and regulating virus development in the insect vector.

2 Vector Species of Culicoides

BTV occurs across the world in a band stretching from about latitude 40° N to 35° S. Within this zone it is found in Africa, the Middle East, Asia, Australia and the Americas. Several incursions have also occurred into Southern Europe. The species of *Culicoides* transmitting BTV in many of these areas have already been identified and these will be discussed in this section. However, in other areas the major vector species have yet to be positively identified.

2.1 Africa and the Middle East

C. imicola (*pallidipennis*) is the major vector of BTV throughout this region and numerous isolates of the virus have been made from this species in South Africa (Du Toit 1944; Wirth and Dyce 1985; Erasmus 1988, personal communication, Kenya (Walker and Davies 1971; Wirth and Dyce 1985), Sudan (Mellor et al. 1984), and Israel (Braverman and Galun 1973; Braverman et al. 1981; Braverman et al. 1985; Shimshony 1987). The range of *C. imicola* in the Afro-Middle Eastern region extends eastwards into the Arabian peninsula (Mellor and Al Busaidy, unpublished observations 1984–1988) and Iran (Navai 1977) and northwards into Turkey (Jennings et al. 1983). Countries within the region that have been shown to support populations of *C. imicola* have invariably reported serological or clinical evidence of bluetongue. In Turkey the presence of *C. imicola* was totally unsuspected until the BTV epizootics in Western Turkey in 1977–1979 (Yonguc et al. 1982; Yonguc 1987). Subsequently large and widespread populations of *C. imicola* were identified, particularly in those areas where bluetongue disease had occurred (Jennings et al. 1983; Jennings et al. 1989). In the Afro-Middle Eastern region outbreaks of BTV seem to be invariably linked with the presence of *C.imicola*.

In Kenya, in addition to *C. imicola*, BTV has been isolated from *C. tororoensis*, a closely related species, and also from *C. milnei* (Walker and Davies 1971). However, these findings have not been repeated and these two species of midge have not been connected with the transmission of BTV elsewhere. *C. obsoletus*, another species of midge closely related to *C. imicola*, and midges of the *C. schultzei* group are also suspected of transmitting BTV in the Afro-Middle Eastern region. BTV has been isolated from *C. obsoletus* in Cyprus (Mellor and Pitzolis 1979) and *C. schultzei* is known to transmit the closely related epizootic haemorrhagic disease (EHD) virus in the Sudan (Mellor et al. 1984a). However, further evidence to link any of these four species of midge to BTV transmission has not been forthcoming, and it seems likely that they are of only local or minor significance in the epidemiology of BTV in the region.

2.2 Asia

Bluetongue in sheep has been recorded in Pakistan (Sarwar 1962; Howell and Verwoerd 1971; Howell 1963) and India (Sapre 1964; Uppal and Vasundervar 1980; Bhambani and Singh 1968; Taylor 1989, personal communication), while serological evidence of the virus has also been reported from Japan (Miura et al. 1980), Papua New Guinea, Malaysia and Indonesia (Sellers 1981; Miura et al. 1982). However, isolation of a BTV has not been made from any species of *Culicoides* in this region. In view of the extent of the serological evidence for BTV in the area this is probably more a reflection of research priorities than of any inherent difficulties in isolating BTV from vector midges. In fact several species of *Culicoides* (*C. wadai*, *C. fulvus*, *C. brevitarsis* and

C. oxystoma) that are known or suspected BTV vectors in Australia occur widely across Southeast Asia (WIRTH and DYCE 1985; KUROGI et al. 1987, 1989). Also, *C. imicola* the major Afro-Middle Eastern vector, which is thought to be conspecific with *C. brevitarsis* (DEBENHAM 1978), occurs at least as far east as Iran (NAVAI 1977), India (DYCE and WIRTH 1983) and Laos (HOWARTH 1985). Detailed vector competence studies are clearly required throughout the whole of the Asian region to establish the precise identity, distribution and prevalance of the major BTV vectors in the area.

2.3 Australia

BTV was first identified in the Northern Territory of Australia in 1977, the original isolate being made from a pool of *Culicoides* collected in 1975. Subsequent to 1977 approximately 50 further isolations of BTV, comprising at least eight different serotypes (1, 3, 9, 15, 16, 20, 21 and 23) have been made, mainly from cattle (ST. GEORGE 1985; GARD 1987). Two of these serotypes have so far been isolated from *Culicoides* species; serotype 20 from a pool of 214 *Culicoides* comprising about 12 different species and serotype 1 from both *C. fulvus* and *C. brevitarsis* (ST. GEORGE and MULLER 1984; STANDFAST et al. 1985). Additionally *C. wadai*, *C. actoni*, *C. peregrinus* and *C. oxystoma* (*schultzei* gp) have been shown to support BTV replication after oral infection in the laboratory, and *C. fulvus* and *C. actoni* have transmitted the virus between sheep, also in the laboratory (CYBINSKI et al. 1980; STANDFAST et al. 1985). *C. wadai*, *C. fulvus* and *C. actoni*

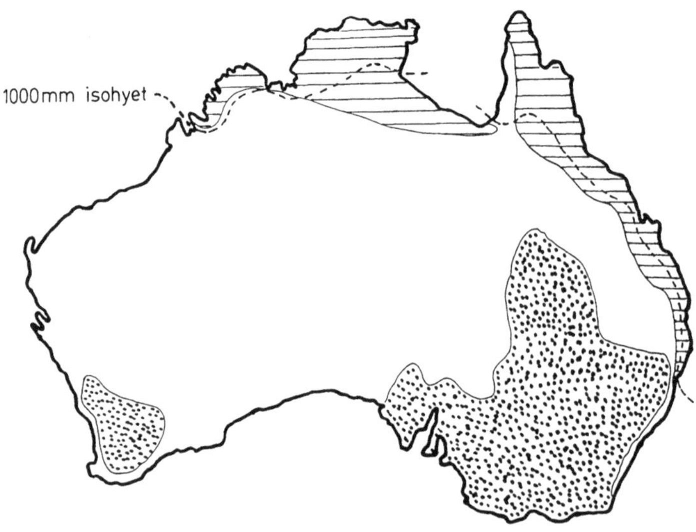

Fig. 1. Distribution of *Culicoides wadai* (*hatched areas*) in Australia in relation to the 1000-mm isohyet and major sheep-rearing areas (*dotted areas*)

exhibited the highest experimental infection rates when feeding on BTV infected sheep, while *C. brevitarsis* was less susceptible to infection. All of these four species are members of the subgenus *Avaritia* and as such are closely related to *C. imicola*. *C. fulvus* and *C. actoni* have an Australian distribution that is restricted to areas with an annual summer rainfall in excess of 1000 mm (Fig. 1). These two species therefore do not occur in the drier sheep-rearing areas and so are unable to transmit BTV to this highly susceptible vertebrate host. *C. wadai*, while initially being similarly restricted in distribution, has been extending its range on an annual basis and now occurs as far south as northern New South Wales (Fig. 1). How much further this highly efficient BTV vector will be able to spread remains to be seen, but its distribution is now verging on some of the major sheep-rearing areas (Ralph 1987). The consequences of this may become apparent over the next few years.

2.4 The Americas

The major vector of BTV throughout the USA and also in the Okanagan Valley in Canada is *C. variipennis*. The virus has been isolated from field collections of this species of midge on numerous occasions (Sellers 1981; Jones et al. 1981). However, *C. variipennis* does not occur in Southern Florida, the Caribbean region, most of Central America and all of South America areas where BTV does occur (Sellers 1981; Homan et al. 1985; Walton et al. 1984; Gibbs and Greiner 1983). In these areas there are several other candidate vectors for BTV, the most likely being *C. insignis* and *C. pusillus* (Greiner et al. 1984; Walton et al. 1984). Partial confirmation of this occurred in 1982 when BTV serotype 2 was isolated from a pool of *C. insignis* in Florida (Greiner et al. 1985). More recently a second BTV serotype has been isolated from *C. insignis* in French Guyana (Lefevre 1988, personal communication).

Vector competence studies with other New World species of *Culicoides* indicate that *C. debilipalpis* and *C. venustus* are capable of oral infection with BTV. However, even under ideal conditions experimental infection rates were very low ($< 1.9\%$ for *C. debilipalpis*; $< 0.7\%$ for *C. venustus*), which suggests that these species would be inefficient virus vectors in the field (Jones et al. 1983; Mullen et al. 1985).

2.5 Europe

BTV incursions into Spain and Portugal occurred between 1956 and 1960 (Mellor et al. 1983) and into the Greek Islands of Lesbos and Rhodes, in 1979 and 1980 respectively (Vassalos 1980; Dragonas 1981). At the time of these outbreaks there was no information on the identity of the virus vectors in any of these countries. However in 1981, *C. imicola* the major Old World BTV vector was recorded at several locations on the Turkish mainland adjacent to the affected Greek Islands (JENNINGS et al. 1983). Following this, in 1982, *C. imicola* was found to be present in large numbers on Lesbos (BOORMAN and WILKINSON 1983) and in 1984 it was also recorded on Rhodes (BOORMAN 1988). Furthermore, in 1982 and 1984 *C. imicola* was found at several sites in Spain and Portugal (MELLOR et al. 1983; MELLOR et al. 1985), and since that time it has been taken repeatedly in Spain at numerous locations as far north as Madrid (BONED and MELLOR, 1987–1989, unpublished data). The distribution of *C. imicola* in the Mediterranean basin is now known to be almost identical to that of the BTV incursions in the same area (MELLOR 1987). This suggests that *C. imicola* is the only major vector of BTV in this region. Other species of *Culicoides* such as *C. schultzei* (MELLOR et al. 1984b), which is a suspect vector in Australia (STANDFAST et al. 1985); *C. obsoletus*, from which BTV has been isolated in Cyprus (MELLOR and PITZOLIS 1979); and *C. nubeculosus*, which has been shown to be a laboratory vector of BTV (JENNINGS and MELLOR 1988) together have much wider distributions in the Mediterranean basin than *C. imicola.* However, these three species appear to have been of little or no significance during the outbreaks of bluetongue disease in Europe (MELLOR 1987). Nevertheless, since *C. schultzei, C. obsoletus* and *C. nubeculosus* have all been shown to be susceptible to BTV infection they should each be regarded as potential vectors during any future epizootics, particularly if an alternative vector is not immediately apparent.

3 Infection of *Culicoides* with Bluetongue Virus

Individual *Culicoides* are infected with BTV in the wild by imbibing viraemic blood from an infected vertebrate host. As far as is known this is the only way in which wild *Culicoides* are able to acquire an infection with this virus. In the laboratory additional means of infection exist using either a range of artificial feeding techniques (RUTLEDGE et al. 1964; MELLOR 1971) or else via parenteral (intrathoracic) inoculation of the virus (MELLOR et al. 1974).

3.1 Parenteral Inoculation

This method of infection usually makes use of fine glass needles and an apparatus similar to the one described by BOORMAN (1975) to introduce virus directly into the haemocoel of the insect, thereby bypassing the gut system. Since haemocoelic fluid bathes all of the organs of the insect host, virus replication is initiated rapidly in susceptible tissues and without the "lag-phase" that is commonly seen after oral infection (Sect. 3.2). When *C. variipennis*, the major North American vector of BTV is inoculated intrathoracically with this virus replication to a level of about 5.0 $\log_{10}$ tissue culture infection dose ($TCID_{50}$) per insect occurs over a 4-day

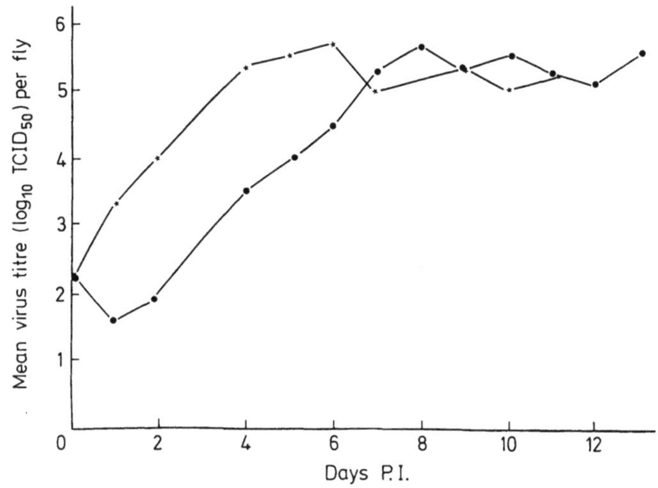

Fig. 2. Replication of bluetongue virus (BTV) in *Culicoides variipennis* after intrathoracic inoculation (*stars*) or oral ingestion (*circles*) of the virus

period (Fig. 2). Of infected individuals 100% develop a persistent infection, and virus transmission may occur after a prepatent period of 4–5 days (FOSTER and JONES 1973; JONES and FOSTER 1966). If *C. variipennis* is orally infected with BTV, maximum virus titers do not develop until at least 7 days post-infection (dpi) and transmission does not normally occur before 10 dpi (Sect. 3.2).

The major disadvantage of inoculation as a means of infection is that in bypassing the gut system of the host and any associated virus barriers (Sect. 4), virus replication may be induced in species and individuals of *Culicoides* that are normally insusceptible per os (MELLOR et al. 1975; MELLOR and JENNINGS 1980). This method will therefore provide data which is potentially misleading should one attempt to extrapolate from parenteral to oral infection as a means of assessing vector competence. However, parenteral inoculation can be of limited value, as a primary screen particularly when dealing with species of *Culicoides* which are reluctant to ingest virus orally, since species which are insusceptible parenterally are highly unlikely to be susceptible orally.

Parenteral infection need not always be regarded as being a totally artificial phenomenon, since it has been shown that double infections of insects with viruses and filarial worms can result in the gut barrier to infection being breached. In such cases the filarial worms act as "natural" inoculators of virus (MELLOR and BOORMAN 1980; TURELL et al. 1984). Insects which are normally insusceptible by the oral route may then support virus replication and transmission may also occur.

3.2 Oral Infection and Replication of BTV in Susceptible *Culicoides*

Female *Culicoides* ingest a wide range of liquid foods including blood, sugars, water and nectar. Most of these liquids are deposited in a blind-ending sac, the mid-gut diverticulum. However, if the food source is blood, contraction of a sphincter muscle at the mouth of the mid-gut diverticulum ensures that most or all of the meal is directed into the hind part of the mid-gut (MEGAHED 1956).

In the wild, susceptible species of *Culicoides* presumably acquire infection with BTV only by imbibing a viraemic blood meal, therefore, the disposition of other foods in the digestive tract is of little importance in regard to virus infection. However, in laboratory studies, the attempted infection of *Culicoides* species with viruses has occasionally taken place using feeding mixtures of virus and sugar solutions, or virus, blood and sugar solutions (BRAVERMAN and SWANEPOEL 1981). The interpretation of results obtained in this way may be difficult since the susceptibility of the acellular tissues of the mid-gut diverticulum is likely to be different from that of the cells of the mid-gut itself. Such a difference has been shown for mosquitoes and Japanese B encephalitis when HALE et al. (1957) were unable to infect *Culex tritaeniorhynchus* with this virus in sugar solution, while similar and lower concentrations of the virus in blood resulted in high infection rates.

Since, under natural conditions, the hind part of the mid-gut of female *Culicoides* receives most or all of the ingested viraemic blood, it is logical to assume that the initial infection with virus occurs in cells in that area (CHAMBERLAIN and SUDIA 1961). Once infection of the mid-gut cells is achieved, then replication ensues, prior to the release of progeny virus into the haemacoel. Secondary target cells, particularly fat body and salivary gland, may then become infected (CHANDLER et al. 1985; BOWNE and JONES 1966).

C. variipennis is the only BTV vector for which detailed virogenic data exist. Fully susceptible individuals of this species support virus replication in a manner similar to that represented schematically in Fig. 2. Female *C. variipennis* are able to ingest approximately 10^{-4} ml blood, so when feeding upon viraemic blood containing 10^6 $TCID_{50}$ virus per ml each midge will ingest about 100 $TCID_{50}$ of virus. This is the day O value. Over the next 24 h the virus titre per individual decreases in what is known as the eclipse or partial eclipse phase. Replication in susceptible tissues then supervenes and virus concentration rises to reach a plateau between days 7 and 9 post-infection, at a level of 5–6 $\log_{10}$ $TCID_{50}$ of virus per midge. This represents a 10^3- to 10^4-fold increase in virus concentration per midge over the day O value. Moreover, this level of virus concentration is maintained for the remainder of the insect's life. Transmission to a vertebrate host becomes possible at 10–14 dpi, subsequent to virus infection of and replication in, the salivary glands (BOWNE and JONES 1966; CHANDLER et al. 1985; FOSTER et al. 1963; LUEDKE et al. 1967; FOSTER and JONES 1973). The titre of virus transmitted during biting by an infected *C. variipennis* has not been accurately estimated but MELLOR (unpublished observations) has recovered < 3 to 20 $TCID_{50}$ of virus after allowing individual BTV-infected *C.variipennis* to feed through a membrane on clean blood. Other authors have also reported that the bite of a single *C. variipennis* is sufficient to infect a susceptible sheep (Foster et al. 1968). When Foster and Jones (1979) infected *C. variipennis* with BTV they observed a complete eclipse phase at 3 dpi when the virus became undetectable in experimental insects. Thereafter, a two-phase virus replication cycle ensued covering days 3 to 4 and 10 to 14 before virus titre reached a plateau at a level of 10^7 egg lethal dose (ELD_{50}) per pool of infected midges. These authors interpreted the initial decrease in virus titre as being due to digestion of the infecting blood meal, with, "attachment, penetration and uncoating of BTV", in the mid-gut cells of infected midges. They suggested that the first increase in virus titre (days 3–4) might correspond to BTV growth "through" the gut wall and that the second increase (days 10–14) could be due to further cycles of virus multiplication in the salivary glands and secondary target organs. Transmission became possible from 14 dpi. The plateau region beyond 14 dpi was conceived as representing either, "a cessation of virus multiplication with a retention of infectivity" or "a steady state of virus replication with its corresponding inactivation". No explanation was advanced to explain why BTV infectivity is apparently limited to a level of approximately 10^7 ELD_{50} of virus per midge. However, the overall controlling factor may be merely the number of susceptible cells available within each infected midge. In this event, and since arboviruses rarely cause the catastrophic damage to susceptible insect cells that they do to mammalian cells, it might be expected that once all susceptible cells are infected, virus replication would persist in them for the duration of their life span. The plateau region recorded by Foster and Jones would therefore be more likely to represent a steady state of virus replication and inactivation rather than a cessation of virus replication.

The infection rate (IR) of *C. variipennis* by the oral route has been shown to be dependent upon the concentration of virus in the blood meal (Jones and Foster 1971a). By feeding colonized *C. variipennis* a series of infected blood meals containing low titres of BTV (3×10^5 ELD_{50}/ml) these authors progressively increased the IR (percentage of individuals actually infected) to a level where it equalled the susceptibility rate (SR, i.e., the percentage of individuals in a population able to be infected). It has also been shown that a single blood meal containing 3×10^6 ELD_{50} of virus per ml will produce an IR which is equal to the SR.

The SR of any vector species of arthropod for an arbovirus rarely equals 100% and the susceptibility of *C. variipennis* for BTV is no exception. Jones and Foster (1978a) have shown not only that different field populations of *C. variipennis* have different susceptibility for several serotypes of BTV, but that a single population could be differently susceptible to different BTV serotypes. Jennings and Mellor (1987) found that even within an established laboratory colony of *C. variipennis*, the response to oral infection with a single serotype of BTV could vary widely between experiments and they recorded IRs ranging from 0% to 51.6%.

JONES and FOSTER (1974) suggested that the oral susceptibility of *C. variipennis* for BTV seems to be controlled by two distinct genetic mechanisms. The first is under the control of a single gene with a dominant allele for resistance. Completely susceptible (100% SR) and highly refractory (0%–3% SR) populations of *C. variipennis* were derived by selective breeding over the course of a single generation from a parent population with a mean SR of about 30%. This suggests a simple form of Mendelian inheritance. The second mechanism controls the threshold of response of *C. variipennis* to BTV. It apparently operates independently of the first mechanism and depends upon the concentration of virus in the blood meal (JONES and FOSTER 1974). The exact nature of this second genetic mechanism is difficult to determine, but it appears to be similar to the proposed viral modulating gene(s) for Western equine encephalomyelitis in *C. tarsalis* (HARDY et al. 1983; KRAMER et al. 1981).

3.3 Persistence of Infection

Although certain long-lived arthropods such as ticks are known to exhibit a self-clearing mechanism for infecting arboviruses (PLOWRIGHT et al. 1970), it is generally accepted that insects once infected remain so for life (CHAMBERLAIN and SUDIA 1961). However, JONES and FOSTER (1971a) have demonstrated diminished BTV infection rates for *C. variipennis* midges that have imbibed a series of infective blood meals. These authors have theorized that this phenomenon could be due to depletion of virus in infected midges once the insects have attained a certain physiological age. This observation has not yet been confirmed by other workers and the same authors had not apparently detected virus depletion in older midges during earlier studies (JOCHIM and JONES 1966; JONES and FOSTER 1966; FOSTER et al. 1968). It may be that the loss of BTV infection exhibited by some midges is influenced not so much by longevity itself as by other factors such as repeated feeding which might cause the "physiological aging" suggested by JONES and FOSTER. However, in the absence of a phenomenon such as this, *C. variipennis*, once persistently infected with BTV can remain so for periods of at least 35 dpi (FOSTER and JONES 1979) and transmission by some infected individuals may occur up to at least 21 dpi (LUEDKE et al. 1976).

3.4 Oral Susceptibility of *C. Variipennis* to Different Bluetongue Virus Serotypes

Individual populations of *C. variipennis* in the field may have widely varying SRs for different BTV serotypes (JONES and FOSTER 1978a, 1979). Some populations have been shown to have SRs varying from 0% to 21% for different BTV serotypes while other populations exhibited rates that varied between 16% and 69% (JONES and FOSTER 1978a). SRs within single field populations of *C. variipennis* also seem to be highly variable with time, and changes from 0% to 13% have been recorded in consecutive years. Usually a population of *C. variipennis* is most susceptible to strains of the BTV serotype that is circulating in that population at the time and is less susceptible to all other BTV serotypes (JONES and FOSTER 1978a). It is obviously important when dealing with vector populations which may exhibit a heterogeneous response to infection with a single BTV serotype or between different serotypes to use as large a sample of insects for each test as possible. Limited testing using small batches of insects is clearly not a reliable method of estimating infection or susceptibility rates under these conditions, since it may identify only part of the response of a vector population. It is therefore inadvisable to attempt to define the susceptibility of any vector population, particularly an apparently refractory one, without extensive testing. Sample sizes of at least 100 insects per test are recommended (JENNINGS and MELLOR 1987).

4 Barriers to the Infection of Arthropods with Viruses

Over 50 years ago STOREY (1933) demonstrated that if the integrity of the mesenteron of a leaf-hopper (*Circadulina mbila*) was disrupted by puncture with a needle, strains of the insect that previously would not transmit maize-streak virus became transmitters. This suggested to him that the mesenteron itself could provide a barrier to the transmission of maize-streak virus. Subsequent to STOREY's original observations demonstration of the same phenomenon with mosquitoes and arboviruses has been extensively documented. MERRILL and TENBROECK (1935) showed that *Aedes aegypti* could not transmit Eastern equine encephalomyelitis (EEE) virus after oral ingestion of the virus. However, if the mesenteron was punctured immediately after engorgement of an infected blood meal, the mosquito could transmit the virus, MCLEAN (1953) demonstrated that Murray Valley encephalitis (MVE) virus replicated in *Culex annulirostris* after ingestion or intrathoracic inoculation of virus, whereas virus replicated in *Anopheles annulipes* only after inoculation. Furthermore, he found that after inoculation, virus replication occurred in cells of the mesenteron of *Cx. annulirostris* but not in those of *An. annulipes*, clearly demonstrating that viral susceptibility per os was determined at the level of the mesenteron.

CHAMBERLAIN and SUDIA (1961), CHAMBERLAIN (1968), MURPHY et al. (1975), MCLINTOCK (1978), and TINSLEY (1975) have outlined several hypotheses to attempt to explain this "gut barrier" to infection of mosquitoes with viruses. However, it has become apparent that the mesenteron is not the only site at which interference with the normal infection and subsequent transmission of an arbovirus by an arthropod may occur. CHAMBERLAIN and SUDIA (1961) showed that *An. quadrimaculatus* is unable to transmit EEE virus even though 70% of females contained a high concentration of virus. MELLOR and WILKINSON (unpublished) have shown that high proportions of some populations of the soft tick *Ornithodoros moubata* can be persistently infected with certain strains of

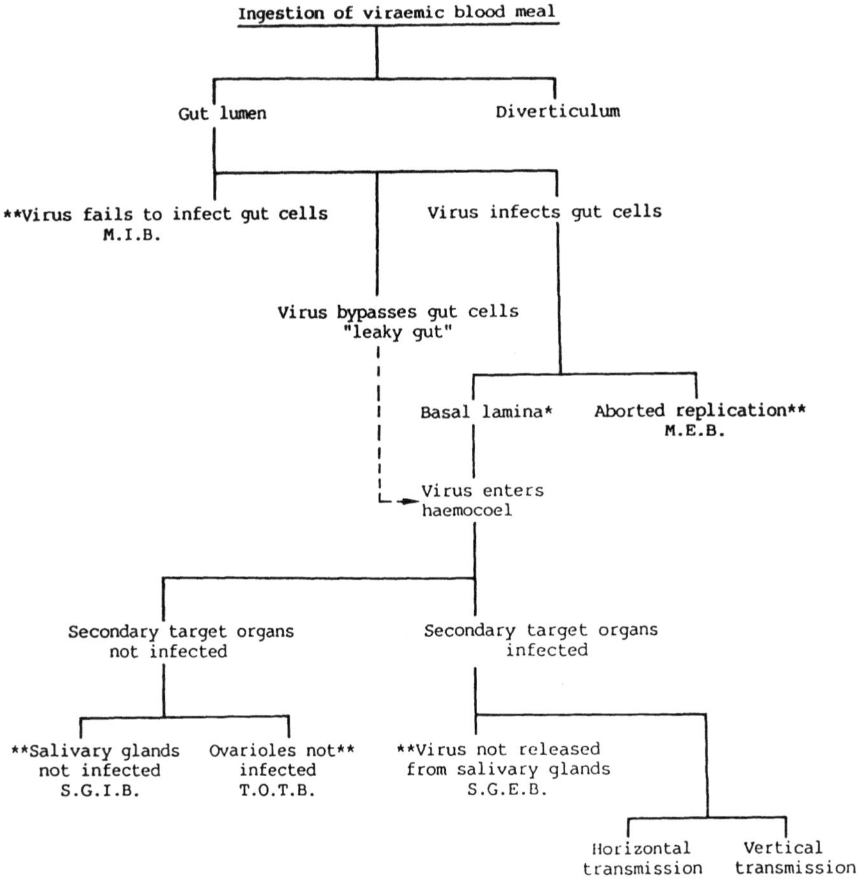

Fig. 3. Hypothesized(*) and conceptual(**) barriers to the arbovirus infection of mosquitoes. M.I.B., mesteron infection barrier; M.E.B., mesenteron escape barrier; S.G.I.B., salivary gland infection barrier; S.G.E.B., salivary gland escape barrier; T.O.T.B., transovarial transmission barrier. (Adapted from HARDY et al. 1983)

African swine fever virus (ASFV), but transmission occurs only rarely or not at all.

It is becoming increasingly evident that vector competence of arthropods for arboviruses is a complex subject and may be associated with multiple barrier systems. Figure 3 depicts a summary of the hypothetical and conceptual barriers to viral infection of mosquitoes from the stage of initial ingestion of an infectious blood meal until the virus is transmitted orally or transovarially (HARDY et al. 1983). Briefly, the major barriers that an arbovirus may have to surmount upon being deposited in the mid-gut of an haematophagous arthropod in order to develop a fully patent infection are:

1. Infection of the mid-gut cells-mesenteron infection barrier
2. Dissemination from the infected gut cells into the haemocoel-mesenteron escape barrier
3. Infection of the salivary glands-salivary gland infection barrier
4. Release from the salivary glands-salivary gland escape barrier.

A further barrier may also exist which prevents virus infection of the ovaries and transmission to the progeny, the transovarial transmission barrier.

HARDY et al. (1983) and MITCHELL (1983) have published comprehensive reviews dealing with all aspects of these and other barriers to the infection of mosquitoes by arboviruses. Unfortunately, even though the principles are likely to be the same, such a wealth of detailed information is not yet available with respect to *Culicoides* and arboviruses.

4.1 Barriers to the Infection of *Culicoides* with Blutongue Virus

Clearly some female *C. variipennis* exhibit a mesenteron infection barrier to BTV. JONES and FOSTER (1974, 1978a, 1979), and JENNINGS and MELLOR (1987) have described how different populations of *C. variipennis* contain a proportion of individuals that are refractory to oral infection with BTV even though 100% of each population can be infected parenterally (JONES and FOSTER 1966; FOSTER and JONES 1973; JOCHIM and JONES 1966). Manifestation of this barrier is apparently controlled by two genetic mechanisms (Sect. 3.2).

JENNINGS and MELLOR (1987) have also demonstrated the presence of a mesenteron escape barrier (MEB) to BTV in *C. variipennis*. They have shown that the maximum level of BTV replication in orally susceptible *C.variipennis* varies from less than 1 to over 5 $\log_{10}$ $TCID_{50}$ of virus per insect.

Persistently infected *C. variipennis* containing less than 2.5 $\log_{10}$ $TCID_{50}$ of virus consistently failed to transmit, whereas virus transmission was regularly demonstrated by midges containing 2.7–5.1 $\log_{10}$ $TCID_{50}$ of virus. Furthermore, MELLOR and JENNINGS (unpublished data 1987) dissected the mid-guts from female *C. variipennis* orally infected with BTV at intervals ranging from 10–14 dpi. In those midges containing less than 2.5 $\log_{10}$ $TCID_{50}$ of BTV, the virus was completely restricted to the mid-gut cells and had failed to disseminate to the secondary target organs, including the salivary glands. These midges, although persistently infected with BTV were therefore incapable of operating as vectors and quite clearly exhibited a MEB. Up to 43.6% of persistently infected *C. variipennis* were found to express such a barrier. This means that should susceptibility to infection be the sole criterion used to determine the BTV vector efficiency of *C. variipennis* populations, there will be a serious risk of over-estimating the vector potential of those populations. The proportion which is actually able to transmit the virus must be determined to avoid such errors.

JONES and FOSTER (1971b) were unable to demonstrate transovarial transmission of BTV by *C. variipennis* when the first to the fifth egg batches of over 440 infected females were reared and tested for the presence of virus. These authors therefore suggested that transovarial transmission of BTV was not likely to occur under normal conditions for *C. variipennis*. JENNINGS (1980, personal communication) also tested over 1000 progeny from BTV-infected female *C. variipennis* and was unable to find any evidence for transovarial transmission of the virus. Neither of these two pieces of work exclude all possibility of transovarial transmission of BTV by *C. variipennis*. However, in the absence of any other evidence in favour of this method of transmission, either experimental or epidemiological, they do suggest that the phenomenon, if it does occur at all, is likely to be rare. Further work is clearly necessary on this subject since studies with other groups of haematophagous Diptera (mosquitoes and phlebotomids) have shown that individuals or different populations of a vector species can vary dramatically in their ability to transmit viruses transovarially (TESH 1981). No information on salivary gland infection barriers and salivary gland escape barriers to BTV is available for any *Culicoides* species.

5 Other Factors Involved in Bluetongue Virus Infection of *Culicoides*

The presence of various barriers to the infection of *Culicoides* with BTV provides a means whereby virus infection of these insects is mediated. However, few studies have been carried out in this area and the nature of the mechanisms controlling these barriers is at the moment poorly understood. In this context some recent studies (MELLOR et al., in press) have suggested that a modification of the "state" of the virus particles themselves by digestive enzymes within the *Culicoides* gut might affect the ability of the particles to infect mid-gut cells. There is no direct evidence concerning the composition of *Culicoides* digestive enzymes, but many groups of other haematophagous insects including tsetse flies, muscids, tabanids, sandflies and mosquitoes are known to secrete mixtures of proteases, especially chymotryspin and trypsin, into the mid-gut (HOUSEMAN 1980; CHAMPLAIN and FISK 1956; THOMAS and GOODING 1976; AKOV 1972; GOODING 1972; SPIRO-KERN and CHEN 1972; and BRIEGEL and LEA 1975). It would be surprising if this is not also the case for *Culicoides*.

MERTENS et al. (1987) have shown that treatment of intact BTV particles with chymotrypsin and trypsin cleaves protein 2 in the outer capsid to give rise to an infectious subviral particle (ISVP). A third particle type (core) can also be produced by the uncoating of either ISVPs or intact virus particles in vitro. ISVPs have a similar specific infectivity for mammalian (BHK 21) cells as do intact, non-aggregated virus particles, while core particles have only very low levels of infectivity for mammalian cells (MERTENS et al. 1987). MELLOR et al. (in press) have now shown that all three particle types are orally infective for BTV-susceptible *Culicoides*.

ISVPs appear to be between 100 and 500 times more infectious for *Culicoides* than do intact virus particles. This finding has since been supported by other studies which have shown that ISVPs are approximately 100 times more infectious than intact virus particles for insect cells (*Ae. albopictus*) when in cell culture (JENNINGS 1989, unpublished observations). Since, therefore, ISVPs of BTV seem to be more infectious for insect cells, it would be advantageous to the virus if intact particles ingested from a vertebrate host by a prospective vector were to be converted into ISVPs in the insect's gut. The presence of proteases such as trypsin and chymotrypsin which are known to occur naturally in the guts of haematophagous insects might well accomplish this. Indeed, since ISVPs seem to be between 100 and 500 times more infective for insects than intact virus particles, the conversion of merely 0.2%–1.0% of ingested intact particles into ISVPs would account for all of the apparent infectivity of the intact virus particles.

Oral infection of *C. variipennis* with core particles of BTV has shown that these particles have a similar level of infectivity to intact virus particles. Since the outer capsid is completely removed from core particles, their infectivity cannot be due to coversion to either of the other two particle types. It therefore seems that core particles are infectious in their own right for vector insects. This suggests that the initial stages of BTV core-insect cell interaction and entry must involve different receptors or different mechanisms to those involving ISVP or virus particles. Also, since core particles are virtually uninfectious for mammalian cells, it also suggests that this mechanism is peculiar to interactions with insect cells.

6 Summary

BTV is maintained in nature by an endless series of alternating cycles of replication in *Culicoides* midges and various mammalian ruminant species. Experimentation has shown that the ability of the virus to infect *Culicoides* persistently and be transmitted by them is restricted to a relatively small number of species. In essence, therefore, the world distribution map of BTV is little more than a distribution map of competent insect vectors.

Once ingested by a competent vector, BTV attaches to the luminal surface of the mid-gut cells, infects these cells and replicates in them. Progeny virus is then released through the basement lamina into the haemocoel from where the secondary target organs including the salivary glands are infected. Subsequent to virus replication in the salivary glands transmission can taken place. The whole cycle from infection to transmission takes between 10–15 days at 25°C and individual vectors once infected usually remain so for life.

Not all female midges within a vector species are susceptible to infection with BTV, or if infected, are competent to transmit the virus. A series of barriers or constraints exists within certain individuals of a vector species which either prevents virus infection or else restricts it in such a way as to stop transmission.

Each population of a vector species of *Culicoides* has a variable proportion of these so-called refractory midges. The refractory and susceptible traits for BTV within a vector species are under genetic control, and by selective breeding, highly susceptible or completely insusceptible populations can be obtained. However, the mechanisms by which these traits are expressed are poorly understood. Further studies are therefore urgently required to determine the precise biochemical nature of these mechanisms and their mode of operation.

References

Akov S (1972) Protein digestion in haematophagous insects. In: Rodriques JG (ed) Insect and mite nutrition. North-Holland, Amsterdam

Bhambani BD, Singh PP (1968) Bluetongue disease of sheep in India. Indian Vet J 45: 370–371

Boorman J (1975) Semi-automatic device for inoculation of small insects with viruses. Lab Pract 24: 90

Boorman J (1986) Presence of bluetongue virus vectors on Rhodes. Vet Rec 118: 21

Boorman J (1988) Taxonomic problems in *Culicoides* of south west Asia, in particular of the Arabian Peninsula. In: Service MW (ed) Biosystematics of haematophagous insects. Clarendon, Oxford, pp 271–282

Boorman JPT, Wilkinson PJ (1983) Potential vectors of bluetongue in Lesbos Greece. Vet. Rec. 113: 395–396

Bowne JG, Jones RN (1966) Observations on bluetongue virus in the salivary glands of an insect vector, *Culicoides variipennis*. Virology 30: 127–133

Braverman Y, Galun R (1973) the occurrence of *Culicoides* in Israel with reference to the incidence of bluetongue. Refuah Vet 30: 121–127

Braverman Y, Swanepoel R (1981) Infection and transmission trials with Nyabira virus in *Aedes aegypti* (*Diptera, Culicidae*) and two species of *Culicoides* (*Diptera, Ceratopogonidae*). Zimbabwe Vet J 12: 13–17

Braverman Y, Rubina M, Frish K (1981) Pathogens of veterinary importance isolated from mosquitoes and biting midges in Israel. Insect Sci Appl 2: 157–161

Braverman Y, Linley JR, Marcus R, Frish K (1985) Seasonal survival and expectation of infective life of *Culicoides* spp. (*Diptera, Ceratopogonidae*) in Israel with implications for bluetongue virus transmission and a comparison of the parous rate in *C. imicola* from Israel and Zimbabwe. J Med Entomol 22: 476–478

Briegal H, Lea AO (1975) Relationships between protein and proteolytic activity in the mid-gut of mosquitoes. J Insect Physiol 21: 1597–1604

Chamberlain RW (1968) Arboviruses, the arthropod-borne animal viruses. Curr Top Microbiol Immunol 42: 38–58

Chamberlain RW, Sudia WD (1961) Mechanism of transmission of viruses by mosquitoes. Ann Rev Entomol 6: 371–390

Champlain RA, Fisk FW (1956) The digestive enzymes of the stable fly, *Stomoxys calcitrans* (L). Ohio J Sci 56: 52–56

Chandler LJ, Ballinger ME, Jones RH, Beaty BJ (1985) The virogenesis of bluetongue virus in *Culicoides variipennis*. In: Barber TL, Jochim MM (eds) Bluetongue and related orbiviruses. Liss, New York, pp 245–253

Cybinski DH, Muller MJ, Squire KRE, Standfast HA, St George TD, Uren MF, Dyce AL (1980) Bluetongue vector studies. CSIRO, Division of Animal Health, Research Reports 1979/80, 19

Debenham ML (1978) An annotated check list and bibliography of Australasian region Ceratopogonidae (*Diptera, Nematocera*). Entomology Monograph No. 1, Australian Government Publishing Service, Canberra, 671

Dragonas PN (1981) Fievre catarrhale du mouton; bluetongue. OIE Monthly Circular 9: 10

Du Toit R (1944) The transmission of bluetongue and horse-sickness by *Culicoides*. Onderstepoort J Vet Sci Anim Ind 19: 7–16

Dyce AL, Wirth WW (1983) Reappraisal of some Indian *Culicoides* species in the subgenus Avaritia (Diptera: Ceratopogonidae). Int J Ent 25: 221–225.

Foster NM, Jones RH (1973) Bluetongue virus transmission with *Culicoides variipennis* via embryonating chicken eggs. J Med Entomol 10: 529–532

Foster NM, Jones RH (1979) Multiplication rate of bluetongue virus in the vector *Culicoides variipennis* (*Diptera: Ceratopogonidae*) infected orally. J Med Ent 15: 302–303

Foster NM, Jones RH, McCrory BR (1963) Preliminary investigations on insect transmission of bluetongue virus in sheep. Am J Vet Res 24: 1195–1200

Foster NM, Jones RH, Luedke AJ (1968) Transmission of attenuated and virulent bluetongue virus with *Culicoides variipennis* infected orally via sheep. Am J Vet Res 29: 275–279

Gard GP (1987) Studies of blutongue virulence and pathogenesis in sheep. In: ISSN Tech Bull No. 103, Dept of Ind & Develop Gov Printer, N Territory, Australia pp 1–58

Gibbs EPJ, Greiner EC (1983) Bluetongue infection and *Culicoides* species associated with livestock in Florida and the Caribbean region. In: Compans RW, Bishop DHL (eds) Double-stranded RNA viruses. Elsevier, New York, pp 375–382

Gooding RH (1972) Digestive processes of haematophagous insects. I. A literature review. Quaest Entomol 8: 5–60

Greiner EC, Garris GI, Rollo RT, Knausenberger WI, Jones JE, Gibbs EPJ (1984) Preliminary studies on the *Culicoides* spp as potential vectors of bluetongue in the Caribbean region. Prev Vet Med 2: 389–399

Greiner EC, Barber TL, Pearson JE, Kramer WL, Gibbs EPJ (1985) Orbiviruses from *Culicoides* in Florida. In: Barber TL, Jochim MM (eds) Bluetongue and related orbiviruses. Liss, New York, pp 195–200

Hale JH, Colless DH, Lim KA (1957) Investigation of the Malaysia form of *Culex tritaeniorhynchus* as a potential vector of Japanese B encephalitis virus on Singapore Island. Ann Trop Med Parasitol 51: 17–25

Hardy JL, Houk GJ, Kramer LD, Reeves WC (1983) Intrinsic factors affecting vector competence of mosquitoes for arboviruses. Annu Rev Entomol 28: 229–262

Homan EJ, Lorbacher DE, Ruiz H, Donato AP, Taylor WP, Yuill TM (1985) A preliminary survey of the epidemiology of bluetongue in Costa Rica and Northern Columbia. J Hyg Camb 94: 357–363

Houseman JG (1980) Anterior mid-gut proteinase inhibitor from *Glossina morsitans morsitans* Westwood (*Diptera: Glossinidae*) and its effects upon tsetse digestive enzymes. Can J Zool 58: 79–87

Howarth FG (1985) Biosystematics of the *Culicoides* of Laos (Diptera: Ceratopogonidae). Int J Ent 27: 1–96

Howell PC (1963) Bluetongue. In: Emerging diseases of animals. FAO Agricultural Studies No. 61, Rome, pp 109–153

Howell PC, Verwoerd DW (1971) Bluetongue virus. In: Hess WR, Howell PC, Verwoerd DW (eds) African swine fever virus, bluetongue virus. Springer, Berlin Heidelberg, New York, pp 35–74 (Virology monographs vol 9)

Jennings DM, Mellor PS (1987) Variation in the responses of *Culicoides variipennis* (Diptera, Ceratopogonidae) to oral infection with bluetongue virus. Arch Virol 95: 177–182

Jennings DM, Mellor PS (1988) The vector potential of British *Culicoides* species for bluetongue virus. Vet Microbiol 17: 1–10

Jennings M, Boorman JPT, Ergun H (1983) *Culicoides* from Western Turkey in relation to bluetongue disease of sheep and cattle. Rev Elev Med Vet Pays Trop 36: 67–70

Jennings DM, Hamblin C, Mellor PS (1989) Control of Akabane disease and surveillance of bluetongue and ephemeral fever in Turkey. FAO report, project no. TUR/86/017

Jochim MM, Jones RH (1966) Multiplication of bluetongue virus in *Culicoides variipennis* following artificial infection. Am J Epidemiol 84: 241–246

Jones RH, Foster NM (1966) The transmission of bluetongue virus to embryonating chicken eggs by *Culicoides variipennis* (*Diptera: Ceratopogonidae*) infected by intrathoracic inoculation. Mosquito News 26: 185–189

Jones RH, Foster NM (1971a) The effect of repeated blood meals infective for bluetongue on the infection rate of *Culicoides variipennis*. J Med Entomol 8: 499–501

Jones RH, Foster NM (1971b) Transovarial transmission of bluetongue virus unlikely for *Culicoides variipennis*. Mosquito News 31: 434–437

Jones RH, Foster NM (1974) Oral infection of *Culicoides variipennis* with bluetongue virus:

development of susceptible and resistant lines from a colony population. J Med Entomol 11: 316–323

Jones RH, Foster NM, (1978a) Heterogeneity of *Culicoides variipennis* field populations to oral infection with bluetongue virus. Am J Trop Med Hyg 27: 178–183

Jones RH, Foster NM (1978b) Relevance of laboratory colonies of the vector in arbovirus research—*Culicoides variipennis* and bluetongue. Am J Trop Med Hyg 27: 168–177

Jones RH, Foster NM (1979) *Culicoides variipennis*: threshold to infection for bluetongue virus. Ann Parasitol Hum Comp 54: 250

Jones RH, Luedke AJ, Walton TE, Metcalf HE (1981) Bluetongue in the United States, an entomological perspective toward control. World Animal Rev 38: 2–8

Jones RH, Schmidemann ET, Foster NM (1983) Vector-competence studies for bluetongue and epizootic haemorrhagic disease viruses with *Culicoides venustus* (Ceratopogonidae). Mosquito News 43: 184–186

Kramer LD, Hardy JL, Presser SB, Houk EJ (1981) Dissemination barriers for western equine encephalomyelitis virus in *Culex tarsalis* infected after ingestion of low viral doses. Am J Trop Med Hyg 30: 190–197

Kurogi H, Akiba K, Inaba Y, Matumoto M (1987) Isolation of Akabane virus from the biting midge *Culicoides oxystoma* in Japan. Vet Microbiol 15: 243–248

Kurogi H, Suzuki T, Akashi H, Ito T, Inaba Y, Matumoto M (1989) Isolation and preliminary characterisation of an orbivirus of the Palyam serogroup from the biting midge *Culicoides oxystoma* in Japan. Vet Microbiol 19: 1–11.

Luedke AJ, Jones RH, Jochim MM (1967) Transmission of bluetongue between sheep and cattle by *Culicoides variipennis*. Am J Vet Res 28: 457–460

Luedke AJ, Jones RH, Jochim MM (1976) Serial cyclic transmission of bluetongue virus in sheep and *Culicoides variipennis*. Cornell Vet 66: 536–550.

McLean DM, (1953) Transmission of Murray Valley encephalitis virus by mosquitos. Aust J Exp Biol Med Sci 31: 481–490

McLintock J (1978) Mosquito virus relationships of American encephalitides. Annu Rev Entomol 23: 17–37

Megahed MM (1956) Anatomy and histology of the alimentary tract of the female of the biting midge *Culicoides nubeculosus* Meigen (Dipt: Heleidae = Cerat.). Parasitology 46: 22–47

Mellor PS (1971) A membrane feeding technique for the infection of *Culicoides nubeculosus* Mg. and *Culicoides variipennis* Coq. with *Onchocerca cervicalis* Rail and Henry. Trans R Soc Trop Med Hyg 65: 199–201

Mellor PS (1987) Factors limiting the natural spread of bluetongue virus in Europe. In: Taylor WP (ed) Bluetongue in the Mediterranean region. Commission of the European Communities, Luxembourg, pp 69–74

Mellor PS, Boorman J (1980) Multiplication of bluetongue virus in *Culicoides nubeculosus* Mg. simultaneously infected with the virus and the microfilariae of *Onchocerca cervicalis* R & H. Ann Trop Med Parasitol 74: 463–469

Mellor PS, Jennings M (1980) Replication of Eubenangee virus in *Culicoides nubeculosus* Mg. and *Culicoides variipennis* Coq. Arch Virol 63: 203–208

Mellor PS, Pitzolis G (1979) Observations on breeding sites and light-trap collections of *Culicoides* during an outbreak of bluetongue in Cyprus. Bull Entomol Res 69: 229–234.

Mellor PS, Boorman J, Loke R (1974) The multiplication of Main Drain virus in two species of *Culicoides (Diptera, Ceratopogonidae)*. Arch Ges Virus Forsch 46: 105–110

Mellor PS, Boorman J, Jennings M (1975) The multiplication of African horse sickness virus in two species of *Culicoides* (Dipt Cerat). Arch Virol 47: 351–356

Mellor PS, Boorman JPT, Wilkinson PJ, Martinez-Gomez F (1983) Potential vectors of bluetongue and African horse sickness viruses in Spain. Vet Rec 112: 229–230

Mellor PS, Osborne R, Jennings DM (1984a) Isolation of bluetongue and related viruses from *Culicoides* species in the Sudan. J Hyg Camb 93: 621–628

Mellor PS, Jennings M, Boorman JPT (1984b) *Culicoides* from Greece in relation to the spread of bluetongue virus. Rev Elev Med Vet Pays Trop 37: 286–289

Mellor PS, Jennings DM, Wilkinson PJ, Boorman JPT (1985) *Culicoides imicola*: bluetongue virus vector in Spain and Portugal. Vet Rec 116: 589–590

Merrill MH, Tenbroeck C (1935) The transmission of equine encephalomyelitis virus by *Aedes aegypti*. J Exp Med 62: 687–695

Mertens PPC, Burroughs JN, Anderson J (1987) Purification and properties of virus particles, infectious sub-viral particles, and cores of bluetongue virus serotypes 1 and 4. Virology 157: 375–386

Mitchell CJ (1983) Mosquito vector competence and arboviruses. In: Harris KF (ed) Current topics in vector research, vol 1. Praeger, New York, pp 63–92
Miura Y, Inaba Y, Hayashi S, Takahashi E, Matumoto M (1980) A survey of antibodies to arthropod-borne viruses in Japanese cattle. Vet Microbiol 5: 277–282
Miura Y, Inaba Y, Tsuda T, Tokuhisa S, Sato K, Akashi H (1982) Seroepizootiological survey on bluetongue virus infection in cattle in Japan. Nat Inst Anim Health Q (Jpn) 22: 154–158
Mullen GR, Jones RH, Braverman Y, Husbaum KE (1985) Laboratory infections of *Culicoides debilipalpis* and *C. stellifer* (*Diptera: Ceratopogonidae*) with bluetongue virus. In: Barber TL, Jochim MM (eds) Bluetongue and related orbiviruses. Liss, New York, pp 239–243
Murphy FA, Whitfield SG, Sudia WD, Chamberlain RW (1975) Interactions of vector with vertebrate pathogenic viruses. In: Maramorosch K, Shope RE (eds) Invertebrate immunity. Academic, New York, pp 25–48
Navai S (1971) Midges of the genus *Culicoides* (*Diptera Ceratopogonidae*) from South-West Asia. Ph.D. thesis, University of Maryland
Plowright W, Perry CT, Pierce MA, Parker J (1970) Experimental infection of the argasid tick, *Ornithodoros moubata porcinus*, with African swine fever virus. Arch Ges Virusforsch 31: 33–50
Ralph W (1987) The Australian bluetongue complex. Rural Res 135: 20–26
Rutledge LC, Ward RA, Gould DJ (1964) Studies on the feeding response of mosquitoes to nutritive solutions in a new membrane feeder. Mosquito News 24: 407–419
Sapre SN (1964) An outbreak of "bluetongue" in goats and sheep. Vet Rev 15: 78–80
Sarwar MM (1962) A note on bluetongue in sheep in West Pakistan. Pakistan J Anim Sci 1: 1–2
Sellers RF (1981) Bluetongue and related diseases. In: Gibbs EPJ (ed) Virus diseases of food animals, vol 2. Academic, London pp 567–584
Shimshony A (1987) Bluetongue activity in Israel, 1950–1985: The disease, virus prevalence, control methods. In: Taylor WP (ed) Bluetongue in the Mediterranean region. Commission of the European Communities, Luxembourg, pp 1–22
Spiro-Kern A, Chen PS (1972) Über die Proteasen der Stechmücke *Culex pipiens*. Res Suisse de Zool T 79: 1151–1159
Standfast HA, Dyce AL, Muller MJ (1985) Vectors of bluetongue virus in Australia. In: Barber TL, Jochim MM (eds) Bluetongue and related orbiviruses. Liss, New York, pp 177–186
St George TD (1985) the search for bluetongue viruses in Australia. In: Barber TL, Jochim MM (eds) Bluetongue and related orbiviruses. Liss, New York, pp 295–305
St George TD, Muller MJ (1984) The isolation of a bluetongue virus from *Culicoides brevitarsis*. Aust Vet J 61: 95
Storey HH (1933) Investigations of the mechanism of the transmission of plant viruses by insect vectors. Proc R Soc Lond [Biol] 113: 463–485
Tesh RB (1981) Vertical transmission of arthropod-borne viruses of vertebrates. In: Mckelvey JJ, Eldridge BF, Maramorosch K (eds) Vectors of disease agents, interactions with plants, animals and man. Praeger, New York, pp 122–137
Thomas AW, Gooding RH (1976) Digestive processes of haematophagous insects VIII. Estimation of meal size and demonstration of trypsin in horse flies and deer flies (*Diptera: Tabanidae*). J. Med. Entomol. 13: 131–136
Tinsley TW (1975) Factors affecting virus infection of insect gut tissue. In: Maramorosch K, Shope RE (eds) Invertebrate immunity. Academic, New York, pp 55–63
Turell MJ, Rossignol PA, Spielmore A, Rossi CA, Bailey CC (1984) Enhanced arboviral transmission by mosquitoes that concurrently ingest microfilariae. Science 225: 1039–1041
Uppal PK, Vasudevan B (1980) Occurrence of bluetongue in India. J Comp Microbiol Immunol Infect Dis 1: 18–20
Vassalos M (1980) Cas de fiévre catarrhale due mouton dans l'Ile de Lesbos (Gréce). Bull Off Int Epiz 92: 547–555
Walker A, Davies FG (1971) A preliminary survey of the epidemiology of bluetongue in Kenya. J Hyg Camb 69: 47–61
Walton TE, Barber TL, Jones RH, Luedke AJ (1984) Epizootiology of bluetongue virus: transmission cycle, vectors and serotypic distribution in the Americas. Prev Vet Med 2: 379–388
Wirth WW, Dyce AL (1985) The taxonomic status of the *Culicoides* vectors of bluetongue viruses. In: Barber TL, Jochim MM (eds) Bluetongue and related orbiviruses. Liss, New York, pp 151–164
Yonguc, AD (1987) Bluetongue in Turkey. In: Taylor WP (ed) Bluetongue in the Mediterranean region. Commission of the European Communities, Luxembourg, pp 23–26
Yonguc AD, Taylor WP, Csontos L, Worrall E (1982) Bluetongue in Western Turkey. Vet Rec 111: 144–146

Immune Response to Bluetongue Virus Infection

J. L. STOTT[1] and B. I. OSBURN[2]

1 Introduction 163
2 Immunological Techniques 164
2.1 Characterization of Humoral Immune Responses 164
2.1.1 Serotype-Specific Antibody 164
2.1.2 Serogroup-Specific Antibody 165
2.2 Characterization of Cellular Immune Responses 166
3 Immune Response of Sheep. 166
3.1 Humoral Immune Responses 166
3.1.1 Serogroup Specific Antibody 166
3.1.2 Serotype-Specific Antibody 166
3.1.3 Protective Immunity 167
3.1.4 Virus Clearance 168
3.2 Cell-Mediated Immune Responses 169
3.2.1 Protective Immunity 169
3.2.2 Virus Clearance 170
3.3 Fetal Immune Responses. 170
4 Immune Response of Cattle 170
4.1 Humoral Immune Responses 170
4.1.1 Serogroup-Specific Antibody 170
4.1.2 Serotype-Specific Antibody 171
4.1.3 Protective Immunity 171
4.1.4 Virus Clearance 171
4.2 Cell-Mediated Immune Responses 172
4.2.1 Protective Immunity/Virus Clearance 172
4.3 Immune-Mediated Pathogenesis 172
4.4 Fetal Immune Responses. 173
5 Immune Responses of Mice 173
5.1 Humoral Immune Response. 174
5.2 Cell-Mediated Immune Responses 174
References 175

1 Introduction

Immune responses to bluetongue virus (BTV) have been studied in a variety of animal species with the majority of studies being on the natural ruminant hosts, sheep and cattle. Additional studies have been done on murine species for the

[1] Dept. Microbiology & Immunology, School of Veterinary Medicine, Davis, CA 95616, USA
[2] Assoc. Dean, Research & Graduate Education, School of Veterinary Medicine, Davis, CA 95616 USA

Current Topics in Microbiology and Immunology, Vol. 162

purpose of developing monoclonal antibodies and as a potential model system for elucidation of protective immune responses. Taken together, these studies have adequately identified VP2 as the primary, and possibly the only, viral protein responsible for inducing virus-neutralizing antibody. Characterization of cellular immune responses, specifically cytotoxic T lymphocytes, has been limited to demonstration of their existence and permissive specificity relative to BTV serotype. However, no information on their protein or peptide specificities has been reported. Studies directed at associating specific immune responses, cellular versus humoral, with protective immunity are for the most part contradictory and/or inconclusive. Elucidation of the specific immune responses, and their protein specificities, responsible for inducing protective immunity and virus clearance will require additional studies employing new approaches. This chapter describes the immunological techniques employed and reports on the immune responses which have been elicited in sheep, cattle, and mice.

2 Immunological Techniques

2.1 Characterization of Humoral Immune Responses

In vivo systems, specifically ovine infectivity studies, initially identified the plurality of BTV strains through cross-challenge experiments (NEITZ 1948). Such techniques were replaced by in vitro virus-neutralization assays which have been responsible for the identification of at least 24 BTV serotypes (HAIG et al. 1956; HOWELL 1960, 1970). All subsequent studies directed at the development of subtype, serotype, and serogroup assays have relied upon the virus-neutralization test to establish BTV specificity. Serological techniques used to study humoral immune responses to BTV can be divided into those that are serotype specific and those that are serogroup specific.

2.1.1 Serotype-Specific Antibody

Virus-neutralization assays in cell culture have been the most popular in vitro technique for identification of BTV serotype. Such assays have typically employed a plaque-reduction technique or alternatively measured neutralizing activity in the fluid phase by inhibition of cytopathic effect as determined by visual observation or by staining with crystal violet of the cell monolayer (BARBER and JOCHIM 1976; DELLA PORTA et al. 1981; HAIG et al. 1956; HOWELL et al. 1970; MACLACHLAN and THOMPSON 1985; PARKER et al. 1975; THOMAS and TRAINER 1970). An additional technique for identifying serotype-specific antibody has been the hemagglutination inhibition (HI) assay (BLUE et al. 1974; VAN DER WALT 1980). In vivo measurement of virus-neutralizing antibody has been limited to passive protection by monoclonal antibody in sheep and mice after live virus challenge (JEGGO et al. 1984b; LETCHWORTH and APPLETON 1983b).

The viral protein specificity of both neutralizing and HI antibody appears to be predominantly, if not exclusively, directed at epitopes on the viral coat protein, VP2 (HUISMANS and ERASMUS 1981; HUISMANS et al. 1987). However, additional studies have suggested that epitopes on VP3 (WHITE et al. 1985) and VP5 (COWLEY and GORMAN 1989; MERTENS et al. 1989) may carry minor neutralization sites. The viral protein specificity (VP2) of virus-neutralizing antibody has been demonstrated by a variety of techniques including immunoprecipitation (GRUBMAN et al. 1983; HUISMANS and ERASMUS 1981; HEIDNER et al. 1988; LETCHWORTH and APPLETON 1983a), immunization of animals with purified VP2 (HUISMANS et al. 1987) and recombinant VP2 (INUMARU and ROY 1987), and use of reassortant viruses (KAHLON et al. 1983). The complement of neutralization-related epitopes on VP2, and their subtype restriction, are poorly documented. Three neutralization epitopes on VP2 from BTV-17 have been identified using monoclonal antibodies; one epitope appears to be carried by all strains of the virus and two epitopes are expressed by a limited number of isolates (LETCHWORTH and APPLETON 1983a). Immunoprecipitation studies have identified a minimum of two distinct neutralizing epitopes on BTV-10 (HEIDNER et al. 1988). A neutralizing epitope is present on VP2 of both BTV-13 and BTV-2 (RISTOW et al. 1988). Through the study of reassortant viruses (BTV-20 and BTV-21), the existence of conformational-dependent neutralization epitopes has been suggested (COWLEY and GORMAN 1989). Reassortant viruses of the segments, coding for VP2 and VP5 (the two major outer coat proteins), express unique neutralization epitopes not expressed on either parental virus.

2.1.2 Serogroup-Specific Antibody

Many techniques have been employed to identify nonneutralizing, serogroup antibodies. For many years the primary assay for the identification of such an antibody was the complement-fixation (CF) test (BOULANGER et al. 1967; ROBERTSON et al. 1965). Because the primary focus of measuring group-specific antibody was for diagnostic and epidemiological purposes, additional assays were developed to provide a more sensitive measure of exposure to the virus. Such techniques include agar gel immunodiffusion (AGID; JOCHIM and CHOW 1969; PEARSON and JOCHIM 1979), indirect enzyme-linked immunosorbent assay (I-ELISA) (HUBSCHLE et al. 1981; POLI et al. 1982; LUNT et al. 1988), competitive ELISA (C-ELISA) (AFSHAR et al. 1987a, b), and hemolysis in gel (HIG) (JOCHIM and JONES, 1980). The apparent specificity of group-specific antibody is VP7, a major core protein (GUMM and NEWMAN 1982; HUISMANS and ERASMUS 1981); however, other structural and nonstructural viral proteins present in the test antigen probably contribute to positive test reactions. Purity of the antigen preparation, and immunological technique employed, can influence the specificity of such assays. Antigenic determinants common to BTV and epizootic hemorrhagic disease (EHD) virus have been identified (HUISMANS and ELS 1979; HUISMANS et al. 1979). Interpretation of BTV serology using AGID or CF is often compromised by such cross-reactivity; however, use of ELISA appears to minimize this problem (LUNT et al. 1988). Immunoprecipitation (HEIDNER et al. 1988; HUISMANS and ERASMUS 1981; RICHARDS et al. 1988) and immunoblotting (ADKISON et al. 1988; MACLACHLAN et al. 1987) techniques have gained popularity in studies of humoral immune responses since protein specificities can be unequivocally determined.

2.2 Characterization of Cellular Immune Responses

The BTV proteins responible for the mediation of cellular immune responses induced by BTV infection of sheep and mice are not known. Techniques for identification of such immune responses have included lymphocyte blastogenesis (GHALIB et al. 1985; STOTT et al. 1985a), cytotoxicity assays (^{51}Cr release) (JEGGO and WARDLEY 1982c), and transfer of protective immunity via lymphocyte populations enriched for T cells (JEGGO et al. 1984a).

3 Immune Response of Sheep

3.1 Humoral Immune Responses

3.1.1 Serogroup-Specific Antibody

Following infection with BTV, the first appearance, titer, and persistence of both group-specific and type-specific antibodies in sheep is dependent in part on assay sensitivity, individual animal and breed variability, and the virus strain employed. Group-specific antibodies, as determined by AGID, CF, ELISA, immunoprecipitation, and immunoblot, can be identified within the first or second week following virus infection; C-ELISA would appear to be most sensitive by identifying virus-specific antibody within 9 days postinfection (dpi) (AFSHAR et al. 1987a, b). As demonstrated by immunoblot, such antibodies can be directed at all viral structural and nonstructural proteins, including nonneutralizing epitopes on VP2 (ADKINSON et al. 1988). However, the detection of such antibody may vary depending on the type of animals. Similar observations have been made by immunoprecipitation (RICHARDS et al. 1988).

3.1.2 Serotype-Specific Antibody

Development of virus-neutralizing antibody is routinely detected within 2 weeks after infection. Inability to mount a neutralizing-antibody response has never been reported in animals that developed demonstrable viremia. Duration of the humoral immune response is poorly characterized. Such studies would require housing animals in insect-secure facilities for extended periods of time following infection.

3.1.3 Protective Immunity

Prior to discussing the immune response(s) associated with protective immunity, it is appropriate to address the practical problems encountered in the interpretation of results. As described above, many studies directed at assessing protective immunity have relied upon the presence or absence of clinical disease. Such an approach can often be difficult, if not impossible, due to breed and individual animal susceptibility to infection and/or immune responsiveness, environmental stress factors, prior exposure to BTV or related viruses not detected by routine serological techniques, and virus virulence (HOWELL and VERWOERD 1971; HOWELL et al. 1970; LUEDKE and JOCHIM 1968a, b; NEITZ 1948; STOTT et al. 1985a). The inability to reproduce clinical disease routinely in sheep (pyrexia, lesions, and leukopenia) following virus inoculation has not only hampered attempts to associate immune responses with protective immunity but has also complicated assessment of virus attenuation for potential vaccine use. Because of such problems, many investigators currently use development, titer, and duration of postchallenge viremia as a primary correlate of protective immunity.

Studies attempting to associate immune responses, humoral and/or cellular, with protective immunity are rather inconclusive, and often contradictory. Early work by DuTiot (cited by HOWELL and VERWOERD 1971), utilizing challenge experiments to assess the duration of immunity in sheep, suggested short-lived immunity. NEITZ (1948) subsequently demonstrated that this poor immunity could be explained by the existence of heterologous strains of BTV. Furthermore, NEITZ (1948) reported that although these various virus strains exhibited a variable degree of common immunity, it was insufficient to protect against infection with a heterologous virus. Development of in vitro virus neutralization assays facilitated the classification of this heterogeneity into serotypes.

No positive correlation between the development of group-specific antibodies (CF and AGID predominantly) and protective immunity and/or virus clearance has been reported, with the exception that such antibody may augment protection in the presence of passively transferred immune T lymphocytes (JEGGO et al. 1984a). A potential role for nonneutralizing antibody in antibody-dependent cell-mediated lysis or complement-mediated lysis of infected cells has not been demonstrated (JEGGO et al. 1983b), though such studies have been very limited. In order to study this further, the identification of cell-surface viral antigens using antibody probes is necessary.

Association of virus-neutralizing antibody and protective immunity to the homologous virus challenge is good (HOWELL and VERWOERD 1971; JOCHIM et al. 1965; LUEDKE and JOCHIM 1968a). Studies employing passive transfer of antibody with virus-neutralizing activity have resulted in various degrees of protection. Passive transfer (intravenous) of immune sera specific to BTV-3 afforded complete protection, while passive transfer (colostral) of BTV-2 specific antibody afforded partial protection (JEGGO et al. 1984b). However, in the latter study, colostrum was obtained by lambs suckling a ewe that had not been exposed to the homologous virus for 1 year, and thus failure of complete protection was probably due to low antibody titer. Two studies have been reported using passive transfer of the same BTV-17-neutralizing monoclonal antibody with contrasting results. LETCHWORTH and APPLETON (1983b) reported complete protection to homologous virus challenge as determined by absence of postchallenge viremia and clinical response, while JEGGO et al. (1984b) observed only partial protection. Although similar volumes (0.03% body weight) were administered to the recipient sheep, the two investigators used different assays for determining the virus-neutralizing titer of the ascitic fluid; thus, the basis of the contrasting reports remains unknown.

The efficacy of virus-neutralizing antibody in affording protection from homologous serotype challenge has been further established by HUISMANS et al. (1987). VP2 was dissociated from the BTV-10 virion in a salt- and pH-dependent manner and used to immunize sheep (HUISMANS et al. 1987). Following three injections of the purified VP2, sheep were protected against homologous serotype challenge as determined by clinical response; unfortunately, post challenge viremia was not determined.

For the most part, humoral immunity associated with neutralizing antibody appears to protect only against challenge with the homologous virus. Such serotype-specific protection may be limited to sheep that have only been exposed to one serotype prior to the live virus challenge. This contention is supported by the report of JEGGO et al. (1983a) in which serial infection of sheep with two distinct serotypes leads to the development of heterotypic neutralizing antibody that was associated with protection following exposure of the sheep to additional serotypes not previously encountered by the animal. The authors have noted similar development of broadly serogroup-specific neutralizing antibody following serial inoculations of individual sheep with different serotypes or in animals with unusually high neutralizing activity. It is probable that minor neutralization-related epitopes exist that are common to multiple virus serotypes. This is supported by the recent report of a monoclonal antibody with neutralizing activity to BTV-2 and BTV-13 (RISTOW et al. 1988).

3.1.4 Virus Clearance

A positive correlation between development of virus-neutralizing antibody and a decrease in viremia is typically observed (GHALIB et al. 1985; GROOCOCK et al. 1982; HOWELL and VERWOERD 1971; LUEDKE 1969; RICHARDS et al. 1988; UREN and SQUIRE 1982). However, there is poor correlation between postchallenge neutralization titer and clinical response (LUEDKE 1969; LUEDKE et al. 1964). Furthermore, the continued isolation of virus for several weeks, in the presence of such antibody (ERASMUS 1975; LUEDKE 1969; RICHARDS et al. 1988), suggests that other immune responses might be required for total elimination of the virus. Alternatively, this virus pool may persist in the form of infectious virus-antibody complexes (JOCHIM and JONES 1977) or be inaccessible to antibody due to a close association of virus with erythrocytes or other blood cells (LUEDKE 1970;

RICHARDS et al. 1988). In the latter case, normal turnover of such cells may be required for the release of virus such that antibody can interact and neutralize this persistent pool.

3.2 Cell-Mediated Immune Responses

3.2.1 Protective Immunity

Induction of cell-mediated immune (CMI) responses following BTV infection has been suggested by both indirect and direct evidence. Virus-neutralizing antibody may not be the only immune response capable of affording protective immunity since sheep can resist challenge with live virus in the absence of neutralizing antibody (JOCHIM et al. 1965; LUEDKE and JOCHIM 1968a; STOTT et al. 1985a).

A possible role for CMI response in protective immunity was first suggested by immunity afforded to sheep following immunization with an inactivated virus preparation which did not induce neutralizing antibody (STOTT et al. 1985a). Upon challenge of Warhill sheep with live virus, viremia was absent compared with control animals. The development of postchallenge neutralizing antibody was inversely correlated to protective immunity. The development of a blastogenic response to BTV antigen in a lymphocyte-stimulation assay correlated most closely with protection and was considered to be evidence of a CMI response. It should be noted that similar experiments with Suffolk-cross sheep were not as conclusive since 30% of the animals tested responded to challenge with intensified clinical signs of bluetongue disease. Such disease intensification had been noted following challenge with virulent virus, of sheep previously exposed to low concentration of BTV (JOCHIM et al. 1965). The immunological basis of such sensitization has not been defined. However, immunological responsiveness to BTV antigen varies with different breeds of sheep (BERRY et al. 1982; NEITZ 1948) and may well play a role in the expression of clinical disease.

Recently more definitive in vitro cytotoxicity studies have demonstrated induction of cytotoxic T lymphocytes following infection of sheep with BTV; peak activity was observed 14 days after infection (JEGGO and WARDLEY 1982c). Passive trasnfer of immune thoracic duct lymphocytes, depleted of B cells, to recipient monozygotic twins afforded partial protection to homologous virus challenge as determined by reduced viremia and febrile response (JEGGO and WARDLEY 1986; JEGGO et al. 1984a). Transfer of such immune lymphocytes to recipient animals with preexisting circulation levels of nonneutralizing antibody resulted in total protection following challenge with a heterologous serotype. While these studies employed few animals, they provided the first evidence that CMI responses are induced in sheep following BTV infection, that a population of these cells has potential cross-serotype reactivity, and that such responses can afford at least some degree of protective immunity.

3.2.2 Virus Clearance

No direct evidence of a role for CMI in virus clearance has been reported. The report that peak cytotoxic T-cell activity is observed 14 days postinfection would argue against a role for CTL-mediated virus clearance since viremia may persist for several weeks beyond this time. However, as described for neutralizing antibody, CTL activity could play a major role in the initial reduction of virus titer typically observed 10 to 21 days postinfection.

3.3 Fetal Immune Responses

Because of the demonstrated pathogenicity of BTV for the developing ovine fetus it is appropriate to address the subject of fetal immune response. BTV causes necrosis in the developing nervous system in the 50- to 80-day fetal lamb, which represents a maturational stage when the fetus is not yet immunologically competent.

The fetal lamb is capable of immunologically responding to BTV with neutralizing antibody by 90 days gestation, and is associated with virus clearance (Enright and Osburn 1980; Osburn 1985). Direct inoculation of fetal lambs in the last trimester of gestation does not result in teratogenesis and virus clearance is mediated by immune reponse, as determined by the development of antibody. However, infection of the late-term fetus, or newborn lamb, may result in prolonged viremia (up to 60 days) (Gibbs et al. 1979; Osburn 1985). This prolonged viremia was associated with nondefined parturition-associated immunosuppression; eventual development of neutralizing antibody was associated with viral clearance.

4 Immune Response of Cattle

4.1 Humoral Immune Responses

4.1.1 Serogroup-Specific Antibody

As described for sheep, detection of group-specific antibodies following BTV infection in cattle is typically identified within 1 or 2 weeks postinfection and is dependent upon assay sensitivity. While no association of such group-specific antibodies has been made with protective immunity or virus clearance, their measurement has played a pivotal role in determining prior exposure to BTV for diagnostic, epidemiological, and export purposes (Stott et al. 1983). While AGID is typically used to determine prior exposure of cattle to BTV, the antibody response is often short-lived (MacLachlan et al. 1984a; Pearson et al. 1988). Extensive epidemiological studies in California have demonstrated a lower percentage of seropositive cattle in winter months versus late summer and fall months (STOTT et al. 1981). This information, together with the demonstrated rapid loss of AGID antibody following virus clearance from certain BTV-infected sentinel cattle (STOTT et al. 1985b), would suggest that this antibody is a poor correlate of prior exposure on an individual animal basis. More recent studies indicate that a competitive ELISA (AFSHAR et al. 1987a, b) is a more sensitive indicator of prior exposure to BTV.

The issue of viral protein specificities in sera of cattle previously infected with BTV has been addressed using both immunoprecipitation and immunoblotting. The application of immunoblotting has been limited to the study of experimentally infected colostrum-deprived calves as cross-reaction background makes interpretation of data difficult in other cattle. As for sheep, infection of colostrum-deprived calves with BTV results in the development of antibodies specific for most BTV structural and nonstructural proteins, as determined by immunoblot (MACLACHLAN et al. 1987). Similar studies employing immunoprecipitation techniques have yielded comparable results (RICHARDS et al. 1988).

4.1.2 Serotype-Specific Antibody

Temporal development of virus-neutralizing antibody in BTV-infected cattle typically appears within 1 to 2 weeks postinfection. Duration of this response is poorly characterized.

4.1.3 Protective Immunity

As described for sheep, infection of naive cattle with a single virus serotype results in the development of type-specific neutralizing antibody; such animals are resistant to challenge with the homologous virus serotype. However, following sequential infection of cattle with two different BTV serotypes, heterotypic virus-neutralizing antibody develops and appears to associate with protection against heterologous serotypes not previously encountered (JEGGO et al. 1983b).

4.1.4 Virus Clearance

No correlation has been observed between development of serogroup-specific AGID antibody and clearance of BTV from cattle. Development of virus protein-specific antibody, as determined by immunoprecipitation and/or immunoblot, has also failed to associate such antibodies with virus clearance (MACLACHLAN et al. 1987; RICHARDS et al. 1988).

As described for sheep, the initial development of neutralizing antibody is associated with a rapid drop in virus titer in blood. However, a long-term, low-titer viremia continues to persist for up to 2 months, sometimes longer, in the presence of a high neutralizing-antibody titer (DUTOIT 1962; HEIDNER et al. 1988; LUEDKE et al. 1969; MACLACHLAN and FULLER 1986; MACLACHAN et al. 1987; RICHARDS et al. 1988). The coexistence of virus and neutralizing antibody could be due to the virus being inaccessible to antibody via its close cellular association and/or the presence of infectious or continuously disassociating virus-antibody complexes.

4.2 Cell-Mediated Immune Responses

4.2.1 Protective Immunity/Virus Clearance

CMI responses in cattle have not been defined. However, Stott et al. (1982) demonstrated induction of a BTV antigen-induced lymphocyte blastogenesis and nonneutralizing antibody following immunization of cattle with an inactivated BTV preparation; however, no protective immunity was obvious (Stott et al. 1982). The in vitro lymphocyte proliferative response became minimal following live virus challenge of immunized cattle. This lack of antigen-specific T-lymphocyte blastogenesis could be due to sequestration of immune T cells from the blood into peripheral lymphoid organs and tissues or to the cytopathic infection of BTV antigen-activated lymphocytes. The latter hypotheses would be supported by recent studies directed at defining BTV-T lymphocyte interactions (Stott, unpublished data) in which productive infection of established bovine T-lymphocyte cultures in vitro resulted in cell death.

4.3 Immune-Mediated Pathogenesis

Because of the extremely rare expression of clinical disease in BTV-infected cattle (dermatitis and stomatitis), studies have been directed at defining a possible underlying immunological mechanism. Original observations made by Metcalf et al. (1979), on a major epizootic of clinical bluetongue disease in the southern USA, led them to speculate that clinical bluetongue in cattle was a hypersensitivity reaction induced in certain animals by previous exposure to BTV of a different type or to other related viruses. This speculation is supported by experimental production of a clinical condition that closely paralleled those lesions described for clinical bluetongue in the field. Cattle were given multiple immunizations of inactivated BTV in association with adjuvant (AlOH) and immunopotentiators (levamisol and cimetidine), with subsequent development of nonneutralizing antibody that included BTV-specific IgE (Stott et al. 1982; Anderson et al. 1987). Upon challenge, sensitized animals developed clinical disease (dermatitis and stomatitis) which was associated with elevated levels of IgE, prostaglandins, and histamine (Anderson et al. 1987; Emau et al. 1984). These results were interpreted as being indicative of an immediate type hypersensitivity (type I) and supported by the subsequent description of a natural case of clinical bluetongue in a heifer (Anderson et al. 1985). This heifer exhibited a severe generalized dermatitis, lameness, alopecia, and sloughed the entire muzzle. BTV was isolated from peripheral blood and BTV-specific reagenic antibody was demonstrated using a modified passive cutaneous anaphylaxis (PCA) test. No evidence of bovine virus diarrhea, malignant catarrhal fever, vesicular stomatitis, or photosensitization was identified. Controlled studies using infection with orbiviruses for sensitization, followed by heterologous virus challenge, will be required to establish and characterize an immunological basis for bluetongue disease in cattle.

4.4 Fetal Immune Responses

Natural BTV infection of the developing bovine fetus is well documented (BARNARD and PIENAAR 1976; BROWN and MACLACHLAN 1983; MCKERCHER et al. 1970; RICHARDS et al. 1971; ZUMPT et al. 1978). Experimental studies have established that the outcome of fetal infection is closely tied to gestational period. In vitro exposure of bovine embryos to BTV results in death; however, the embryo is protected from such a lytic effect while it is still within the zona pellucida (BOWEN et al. 1982). Inoculation of fetuses in the second trimester of gestation results in cerebral malformation (BARNARD and PIENAAR 1976; MACLACHLAN and OSBURN 1983; MACLACHLAN et al. 1985; THOMAS et al. 1986), whereas infection late in gestation probably results in the birth of viremic and clinically normal calves (OSBURN and STOTT, unpublished data).

MACLACHLAN et al. (1984a) studied fetal immune responses to BTV following direct inoculation of 125-day-old fetuses. Serum IgM and IgG were identified by 8 and 12 days postinoculation, and BTV group-specific (AGID) and serotype-specific (virus neutralizing) antibody were present at 20 days postinoculation (145 days). Virus persisted in the presence of high titer neutralizing antibody for at least 50 days, but the infection was cleared by birth (150 days postinoculation). Fetal development of a CMI response, as determined by BTV-induced lymphocyte blastogenesis, was never identified, nor did the viral infection appear to compromise the ability of fetal lymphocytes to respond to mitogens (MACLACHLAN et al. 1984b). Regardless of exact mechanisms of virus clearance from the fetus, these studies contrasted with those described earlier by LUEDKE et al. (1977a, b, c) in which calves were born to dams infected with BTV early in gestation. Calves born in these studies exhibited a range of anatomical anomalies, were persistently infected with BTV, and were immunologically unresponsive to the virus. The conflicting nature of these reports is not easily explained. However, immune responses accompany most, if not all, viral infections and true immunological tolerance does not occur (OLDSTONE 1979).

5 Immune Responses of Mice

A brief discussion of murine immune responses is relevant here in that the mouse is the only laboratory animal model that has been used extensively in developing experimental hypotheses relative to specific immune responses, and their protein specificities, associated with protective immunity. Whether the mouse is a realistic model for studying protective immune responses of the natural ruminant hosts of BTV remains to be determined.

5.1 Humoral Immune Response

The study of antibody responses to BTV, and their potential role in contributing to protective immunity, has been confined to the suckling mouse. While mature mice have been immunized for the purpose of developing monoclonal antibodies, they do not express disease and virus replication is probably minimal. Intracerebral inoculation of suckling mice is typically required to induce clinical disease and death. However, some virus isolates have recently been identified that are capable of infecting and killing suckling mice following subcutaneous inoculation (WALDVOGEL et al. 1986). Unpublished experience in our laboratory with BALB/c mice would indicate they must receive virus within approximately 5 days of birth if clinical disease and death is to be observed.

Experiments employing passive transfer (colostral) of BTV group-specific nonneutralizing antibody to suckling mice, followed by intracerebral challenge with homologous and heterologous BTV serotypes, resulted in no protective immunity (STOTT, unpublished data). Such studies would support a nonprotective role of serogroup-specific antibody, as previously described for sheep and cattle.

In contrast to group-reactive antibodies, virus-neutralizing antibodies appear to afford good protective immunity. Passive transfer of monoclonal antibody with virus-neutralizing activity protected suckling mice from homologous virus challenge (LETCHWORTH and APPLETON 1983b). Similarly, newborn mice, suckling immune dams with circulating levels of homologous neutralizing antibodies, are protected. This immunity appears to protect mice against homologous serotype challenge, regardless of virus strain (identified by RNA genome PAGE profiles and virulence for mice by subcutaneous inoculation) (STOTT, unpublished data).

5.2 Cell-Mediated Immune Responses

The induction of BTV-specific, major histocompatibility complex-restricted, cytotoxic T lymphocytes (CTLs) was first demonstrated in mice (JEGGO and WARDLEY 1982a, b, c). In these studies employing ^{51}Cr-release assays, induction of CTLs required replication of the virus, as inactivated virus preparations did not result in generation of cytotoxic activity. Spleen cell preparations derived from mice immunized with a single virus serotype were capable of lysing syngeneic target cells infected with heterologous serotypes, demonstrating their group specificity. These studies would support the nonrestricted nature of bovine CTLs relative to virus serotype.

Acknowledgement. The authors appreciate the invaluable assistance of Dr. N. J. Maclachlan in the preparation of this manuscript.

References

Adkison MA, Stott L, Osburn BI (1988) Temporal development bluetongue virus protein-specific antibody in sheep following natural infection. Vet Microbiol 16: 231–241

Afshar A, Thomas FC, Wright PF, Shapiro JL, Shettigara PT, Anderson J (1987a) Comparison of competitive and indirect enzyme-linked immunosorbent assays for detection of bluetongue virus antibodies in serum and whole blood. J Clin Microbiol 25: 1705–1710

Afshar A, Thomas FC, Wright PF, Shapiro JL, Anderson J, Fulton RW (1987b) Blocking dot-ELISA, using a monoclonal antibody for detection of antibodies to bluetongue virus in bovine and ovine sera. J Virol Methods 18: 271–280

Anderson GA, Kvasnicka WG, Hultine AR (1985) Virus-specific reaginic antibody and clinical disease in a heifer naturally infected with bluetongue virus. Proc Annu Meet Am Assoc Vet Lab Diag 28: 337–346

Anderson GA, Stott JL, Gershwin LJ, Osburn B (1987) Identification of bluetongue virus-specific immunoglobulin E in cattle. J Gen Virol 68: 2509–2514

Barber TL, Jochim MM (1976) Serotyping bluetongue and epizootic hemorrhagic disease virus strains. Proc Annu Meet Am Assoc Vet Lab Diag 18: 149–162

Barnard BJH, Pienaar JG (1976) Bluetongue virus as a cause of hydranencephaly in cattle. Onderstepoort J Vet Res 43: 155–158

Berry LJ, Osburn BI, Stott JL, Farver T, Heron B, Patton W (1982) Inactivated bluetongue virus vaccine in lambs: differential serological responses related to breed. Vet Res Commun 5: 289–293

Blue JL, Dawe DL, Gratzek JB (1974) The use of passive hemagglutination for the detection of bluetongue viral antibodies. Am J Vet Res 35: 139–142

Boulanger P, Ruckerbauer GM, Banister GL, Grap DP, Girard A (1967) Studies on bluetongue. Comparison of two complement-fixation methods. Can J Comp Med 31: 166–170

Bowen RA, Howard TH, Picket BW (1982) Interaction of bluetongue virus with preimplantation embryos from mice and cattle. Am J Vet Res 43: 1907–1911

Brown CC, Maclachlan NJ (1983) Congenital encephalopathy in a calf. Vet Pathol 20: 770–773

Cowley JA, Gorman BM (1989) Cross-neutralization of genetic reassortants of bluetongue virus serotypes 20 and 21. Vet Microbiol 19: 37–51

Della Porta AJ, Herniman KAJ, Sellers RF (1981) A serological comparison of the Australian isolate of bluetongue virus type 20 (CSIRO 19) with bluetongue group viruses. Vet Microbiol 6: 9–21

DuToit RM (1962) The role played by bovines in the transmission of bluetongue in sheep. J S Afr Vet Med Assoc 33: 483–490

Emau P, Giri SN, Anderson GA, Stott JL, Osburn BI (1984) Function of prostaglandins, thromboxane A_2 and histamine in hypersensitivity reaction to experimental bluetongue disease in calves. Am J Vet Res 45: 1852–1857

Enright FM, Osburn BI (1980) Ontogeny of host responses in bovine fetuses infected with bluetongue virus. Am J Vet Res 41: 224–229

Erasmus BJ (1975) Bluetongue in sheep and goats. Aust Vet J 51: 165–170

Ghalib HW, Cherrington JM, Osburn BI (1985) Virological, clinical and serological responses of sheep infected with tissue culture adapted bluetongue virus serotypes 10, 11, 13 and 17. Vet Microbiol 10: 179–188

Gibbs EPJ, Lawman MJP, Herniman KAJ (1979) Preliminary observations on transplacental infection of bluetongue virus in sheep a possible overwintering mechanism. Res Vet Sci 27: 118–120

Groocock CM, Parsonson IM, Campbell CH (1982) Bluetongue virus serotype 20 and 17 infection of sheep: comparison of clinical and serological responses. Vet Microbiol 7: 189–196

Grubman MJ, Appleton JA, Letchworth GJ (1983) Identification of bluetongue virus type 17 genome segments coding for polypeptides associated with virus neutralization and intergroup reactivity. Virology 131: 355–366

Gumm ID, Newman JF (1982) The preparation of purified bluetongue virus group antigen for use as a diagnostic reagent. Arch Virol. 72: 83–93
Haig DA, McKercher DG, Alexander RA (1956) The cytopathogenic action of bluetongue virus on tissue cultures and its application to the detection of antibodies in the serum of sheep. Onderstepoort J Vet Res 27: 171–178
Heidner HW, Maclachlan NJ, Fuller FJ, Richards RG, Whetter LE (1988) Bluetongue virus genome remains stable throughout prolonged infection of cattle. J Gen Virol 69: 2629–2636
Howell PG (1960) A preliminary antigenic classification of strains of bluetongue virus. Onderstepoort J Vet Res 28: 357–363
Howell PG (1970) The antigenic classification and distribution of naturally occurring strains of bluetongue virus. J S Afr Vet Med Assoc 41: 215–223
Howell PG, Verwoerd DW (1971) Bluetongue virus. Virol Monogr 9: 35–74
Howell PG, Verwoerd DW, Oellermann RA (1967) Plaque formation by bluetongue virus. Onderstepoort J Vet Res 34: 317–332
Howell PG, Kumm NA, Botha MJ (1970) The application of improved techniques to the identification of strains of bluetongue virus. Onderstepoort J Vet Res 37: 59–66
Hubschle OJB, Lorenz RJ, Matheka HD (1981) Enzyme-linked immunosorbent assay for detection of bluetongue virus antibodies. Am J Vet Res 42: 61–65
Huismans H, Els HJ (1979) Characterization of the tubules associated with the replication of three different orbiviruses. Virology 92: 397–406
Huismans H, Erasmus BJ (1981) Identification of the serotype-specific and group-specific antigens of bluetongue virus. Onderstepoort J Vet Res 48: 51–58
Huismans H, Bremer CW, Barber TL (1979) The nucleic acid and proteins of epizootic haemorrhagic disease virus. Onderstepoort J Vet Res 46: 95–104
Huismans H, Van Der Walt NT, Cloete M, Erasmus BJ (1987) Isolation of a capsid protein of bluetongue virus that induces a protective immune response in sheep. Virology 157: 172–179
Inumaru S, Roy P (1987) Production and characterization of the neutralization antigen VP2 of bluetongue virus serotype 10 using a baculovirus expression vector. Virology 157: 472–479
Jeggo MH, Wardley RC (1982a) Production of murine cytotoxic T lymphocytes by bluetongue virus following various immunization procedures. Res Vet Sci 33: 212–215
Jeggo H, Wardley RC (1982b) Generation of cross-reactive cytotoxic T lymphocytes following immunization of mice with various bluetongue virus types. Immunology 45: 629–635
Jeggo MH, Wardley RC (1982c) The induction of murine cytotoxic T lymphocytes by bluetongue virus. Arch Virol 71: 197–206
Jeggo MH, Wardley RC (1986) Serial inoculation of sheep with two bluetongue virus types. Res Vet Sci 40: 386–392
Jeggo MH, Gumm ID, Taylor WP (1983a) Clinical and serological response of sheep to serial challenge with different bluetongue virus types. Res Vet Sci 34: 205–211
Jeggo MH, Wardley RC, Taylor WP (1983b) Host response to bluetongue virus. In: Compans RW, Bishop DHL (eds) Double-stranded RNA viruses. Elsevier, New York, pp 353–359
Jeggo MH, Wardley RC, Brownlie J (1984a) A study of the role of cell-mediated immunity in bluetongue virus infection in sheep, using cellular adoptive transfer techniques. Immunology 52: 403–410
Jeggo MH, Wardley RC, Taylor WP (1984b) Role of neutralizing antibody in passive immunity to bluetongue infection. Res Vet Sci 36: 81–85
Jochim MM, Chow TL (1961) Immunodiffusion of bluetongue virus. Am J Vet Res 30: 33–41
Jochim MM, Jones SC (1977) Enhancement of bluetongue and epizootic hemorrhagic disease viral neutralization with anti-gamma globulin. Proc Annu Meet Am Assoc Vet Diag 20: 255–272
Jochim MM, Jones SC (1980) Evaluation of a hemolysis-in-gel test for detection and quantitation of antibodies to bluetongue virus. Am J Vet Res 41: 595–599
Jochim MM, Luedke AJ, Bowne JG (1965) The clinical and immunogenic response of sheep to oral and intradermal administration of bluetongue virus. Am J Vet Res 26: 1254–1260
Kahlon J, Sugiyama K, Roy P (1983) Molecular basis of bluetongue virus neutralization. J Virol 48: 627–632
Letchworth GJ, Appleton JA (1983a) Heterogeneity of neutralization-related epitopes within a bluetongue virus serotype. Virology 124: 300–307
Letchworth GJ, Appleton JA (1983b) Passive protection of mice and sheep against bluetongue virus by a neutralizing monoclonal antibody. Infect Immunol 39: 208–212
Luedke AJ (1969) Bluetongue in sheep: viral assay and viremia. Am J Vet Res 30: 499–509

Luedke AJ (1970) Distribution of virus in blood components during viremia of bluetongue. Proc Annu Meet US Anim Health Assoc 74: 9–21
Luedke AJ, Jochim MM (1968a) Clinical and serological responses in vaccinated sheep given challenge inoculation with isolates of bluetongue virus. Am J Vet Res 29: 841–852
Luedke AJ, Jochim MM (1968b) Bluetongue virus in sheep. Intensification of the clinical response by previous oral administration. Cornell Vet 58: 48–58
Luedke AJ, Bowne JG, Jochim MM, Doyle C (1964) Clinical and pathological features of bluetongue in sheep. Am J Vet Res 25: 963–970
Luedke AJ, Jochim MM, Jones RH (1969) Bluetongue in cattle: viremia. Am J vet Res 30: 511–516
Luedke AJ, Jochim MM, Jones RH (1977a) Bluetongue in cattle: effects of *Culicoides variipennis* transmitted bluetongue virus on pregnant heifers and their calves. Am J Vet Res 38: 1687–1695
Luedke AJ, Jochim MM, Jones RH (1977b) Bluetongue in cattle: effects of vector-transmitted bluetongue virus on calves previously infected in utero. Am J Vet Res 38: 1697–1700
Luedke AJ, Jones RH, Walton TE (1977c) Overwintering mechanism for bluetongue virus: biological recovery of latent virus from a bovine by bites of *Culicoides variipennis*. Am J Trop Med Hyg 26: 313–324
Lunt RA, White JR, Della-Porta AJ (1988) Studies with enzyme-linked immunosorbent assays for the serodiagnosis of bluetongue and epizootic haemorrhagic disease of deer. Vet Microbiol 16: 323–338
Maclachlan NJ, Fuller FJ (1986) Genetic stability in calves of a single strain of bluetongue virus. Am J Vet Res 47: 762–764
Maclachlan NJ, Osburn BI (1983) Bluetongue virus-induced hydranencephaly in calves. Vet Pathol 20: 563–573
Maclachlan NJ, Thompson J (1985) Bluetongue virus-induced interferon in cattle. Am J Vet Res 46: 1238–1241
Maclachlan NJ, Schore CE, Osburn BI (1984a) Antiviral responses of bluetongue virus-inoculated bovine fetuses and their dams. Am J Vet Res 45: 1469–1473
Maclachlan NJ, Schore CE, Osburn BI (1984b) Lymphocyte blastogenesis in bluetongue virus or *Mycobacterium bovis*-inoculated bovine fetuses. Vet Immunol Immunopathol 7: 11–18
Maclachlan NJ, Osburn BI, Ghalib HW, Stott JL (1985) Bluetongue virus-induced encephalopathy in fetal cattle. Vet Pathol 22: 415–417
Maclachlan NJ, Heidner HW, Fuller FJ (1987) Humoral immune response of calves to bluetongue virus infection. Am J Vet Res 48: 1031–1035
McKercher DG, Saito JK, Singh KV (1970) Serologic evidence of an etiologic role for bluetongue virus in hydranencephaly of calves. J Am Vet Med Assoc 156: 1044–1047
Mertens PPC, Pedley S, Cowley J, Burroughs JN, Corteyn AH, Jeggo MH, Jennings DM, Gorman BM (1989) Analysis of the roles of bluetongue virus outer capsid proteins VP2 and VP5 in determination of virus serotype. Virology 170: 561–565
Metcalf HE, Lomme J, Beal VC (1979) Estimate of incidence and direct economic losses due to bluetongue in Mississippi cattle during 1979. Proc Annu Meet US Anim Health Assoc 83: 186–202
Neitz WO (1948) Immunological studies on bluetongue in sheep. Onderstepoort J Vet Sci Anim Indust 23: 93–118
Oldstone MBA (1979) Immune responses, immune tolerance, and viruses. In: Fraenkel-Conrat H, Wagner RR (eds) Comprehensive virology. Plenum, New York, 15: 1–28
Osburn BI (1985) Role of the immune system in bluetongue host-viral interactions. In: Barber TI, Jochim MM (eds) Bluetongue and related orbiviruses. Liss, New York, pp 417–422
Parker J, Herniman KAJ, Gibbs RPJ, Sellers RF (1975) An experimental inactivated vaccine against bluetongue. Vet Rec 96: 284–287
Pearson JE, Jochim MM (1979) Protocol for the immunodiffusion test for bluetongue. Proc Annu Meet Am Assoc Vet Lab Diag 22: 463–471
Pearson JE, Shafer AL, Luedke AJ, Acree JA, Ross GS, Gustafson GA, Peterson LA (1988) Experimental bovine bluetongue infection: serologic response and virus isolation. Proc Annu Meet Am Assoc Vet Lab Diag 31
Poli G, Stott J, Liu YS, Manning JS (1982) Bluetongue virus: comparative evaluation of enzyme-linked immunosorbent assay, immunodiffusion, and serum neutralization for detection of viral antibodies. J Clin Microbiol 15: 159–162
Richards RG, Maclachlan NJ, Heidner HW, Fuller FJ (1988) Comparison of virologic and serologic responses of lambs and calves infected with bluetongue virus serotype 10. Vet Microbiol 18: 233–242

Richards WPC, Crenshaw GL, Bushnell RB (1971) Hydranencephaly of calves associated with natural bluetongue virus infection. Cornell Vet 61: 336–348
Ristow S, Leendersten L, Gorham J, Yilma T (1988) Identification of a neutralizing epitope shared by bluetongue virus serotypes 2 and 13. J Virol 62: 2502–2504
Robertson A, Appeal M, Bannister GL, Ruckerbauer G, Boulanger P (1965) Studies on bluetongue. Complement-fixing activity of ovine and bovine sera. Can J Med Vet Sci 29: 113–117
Stott JL, Else KC, McGowan B, Wilson LK, Osburn BI (1981) Epizootiology of bluetongue virus in Western United States. Proc Annu Meet US Anim Health Assoc 5: 170–180
Stott JL, Anderson GA, Jochim MM, Barber TL, Osburn BI (1982) Clinical expression of bluetongue disease in cattle. Proc Annu Meet US Anim Health Assoc 86: 126–131
Stott JL, Osburn BI, Machlachlan NJ (1983) Diagnosis of bluetongue virus infection in cattle: Virus isolation or serology? Proc Annu Meet Am Assoc Vet Lab Diag 26: 301–318
Stott JL, Barber TL, Osburn BI (1985a) Immunologic response of sheep to inactivated and virulent bluetongue virus. Am J Vet Res 46: 1043–1049
Stott JL, Osburn BI, Bushnell R, Loomis RC, Squire KRE (1985b) Epizootiological study of bluetongue virus infection in California livestock: an overview. In: Barber TI, Jochim MM (eds) Bluetongue and related orbiviruses. Liss, New York, pp 571–582
Thomas FC, Trainer DO (1970) Bluetongue virus: (1) In pregnant white-tailed deer; (2) A plaque reduction neutralization test. J Wildl Dis 6: 384–388
Thomas FC, Randall GCB, Myers DJ (1986) Attempts to establish congenital bluetongue virus infections in calves. Can J Vet Res 50: 280–281
Uren MF, Squire KRE (1982) The clinico-pathological effect of bluetongue virus serotype 20 in sheep. Aust Vet J 58: 1–15
Van Der Walt NT (1980) A haemagglutination and haemagglutination-inhibition test for bluetongue virus. Onderstepoort J Vet Res 47: 113–117
Waldvogel AS, Stott JL, Squire KRE, Osburn BI (1986) Strain-dependent virulence characteristics of bluetongue virus serotype 11. J Gen Virol 67: 765–769
White JR, Breschkin AM, Della-Porta AJ (1985) Immunochemical analyses of Australian bluetongue virus serotypes using monoclonal antibodies. In: Barber TI, Jochim MM (eds) Bluetongue and related orbiviruses. Liss, New York, pp 397–405
Zumpt GF, Bryson RW, Andrews S (1978) Hydranencephaly in calves in the Natal region. J S Afr Vet Med Assoc 49: 32

Appendix

Restriction sites of each BTV gene. E, *EcoR*1; D, *Dde*1; H, *Hinf*1; P, *Pst*1 are given in Fig. 1 and the predicted amino acid sequences of the encoded proteins in Fig. 2.

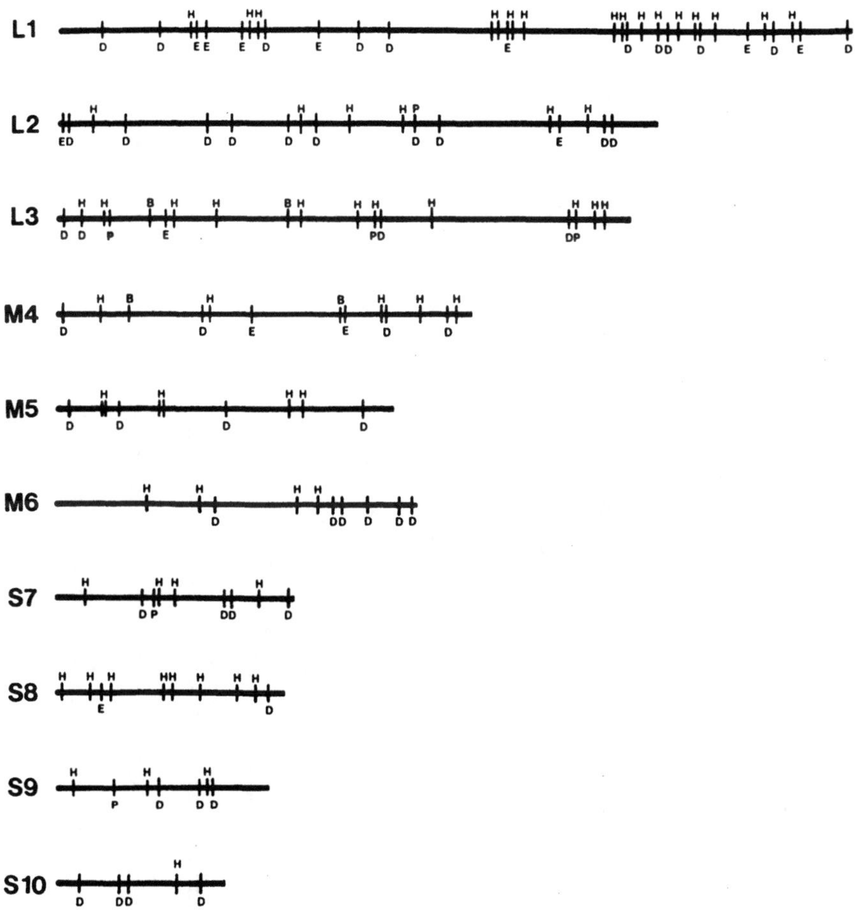

L1

```
           M  V  A  I  T  V  Q  G  A  Q  L  I  K  R  V  V  E  R  F  Y  P  G  I  A  F  D  I  N  E  G  A  C  Y  I  Y  K
GTTAAAATGCAATGGTCGCAATCACCGTGCAAGGTGCACAGCTCATCAAGCGAGTGGTCGAACGCTTTTATCCAGGGATAGCATTTGATATAAATGAAGGAGCATGTTATATATATAAGT
                                                                                                                     120

F  S  D  H  I  R  R  I  R  M  K  H  G  T  K  Y  R  R  Q  A  E  E  I  M  R  S  I  S  L  R  K  E  R  L  Y  G  I  P  V  L
TTTCTGATCATATACGACGTATAAGGATGAAACATGGGACGAAATATCGGCGGCAGGCGGAAGAGATTATGCGCAGTATAAGTCTGAGAAAGGAGCGATTGTATGGGATACCAGTGTTAG
                                                                                                                     240

D  E  V  E  W  K  Y  V  F  D  G  Q  T  F  Q  S  Y  A  F  E  V  Y  V  N  S  I  L  P  W  S  E  L  D  P  E  E  E  F  L  R
ATGAGGTTGAGTGGAAATACGTGTTTGACGGCCAAACGTTCCAAAGTTACGCTTTTGAGGTGTACGTGAACTCAATTTTGCCGTGGAGTGAACTTGATCCGGAGGAGGAGTTCTTACGTA
                                                                                                                     360

N  Y  R  V  S  R  E  T  T  E  V  E  K  F  I  E  F  R  A  K  N  E  M  Q  I  Y  G  D  I  P  I  K  V  W  C  C  F  I  N  E
ATTATAGAGTTTCAAGGGAGACGACTGAAGTGGAAAAATTTATCGAATTTCGTGCTAAAAACGAGATGCAAATATACGGAGATATACCCATTAAAGTATGGTGTTGTTTCATCAACGAAC
                                                                                                                     480

L  S  I  E  L  N  P  I  P  L  G  M  Q  V  M  A  D  F  V  N  R  F  N  S  P  F  H  Q  G  N  R  D  L  S  N  L  E  D  F  Q
TGAGTATAGAATTAAATCCAATCCCTTTAGGGATGCAAGTTATGGCTGACTTCGTAAACCGTTTCAATTCGCCATTCCACCAGGGGAATAGAGATTTATCAAATCTTGAAGATTTTCAAG
                                                                                                                     600

V  A  Y  T  T  P  L  L  F  E  M  C  C  M  E  S  I  L  E  F  N  I  K  M  R  M  R  E  E  D  I  S  A  L  E  F  G  D  I  K
TCGCATATACCACGCCGCTCTTATTTGAGATGTGTTGCATGGAATCAATTTTAGAATTCAATATTAAAATGCGTATGCGTGAAGAAGACATCTCGGCGTTGGAATTCGGTGATATAAAGA
                                                                                                                     720

I  D  P  V  G  L  L  R  E  F  F  I  L  C  L  P  H  P  K  K  I  N  N  V  L  R  A  P  Y  S  W  F  V  K  M  W  G  V  G  A
TTGATCCAGTTGGACTATTGCGTGAGTTCTTTATTCTGTGCTTACCACACCCAAAGAAAATTAACAACGTTTTAAGAGCGCCATATTCTTGGTTTGTGAAGATGTGGGGTGTCGGAGCAG
                                                                                                                     840

D  P  I  V  V  L  Q  S  T  A  G  D  D  R  N  S  K  D  V  F  Y  D  K  F  R  T  E  P  N  R  Y  K  A  L  F  R  S  S  F  Y
ATCCAATCGTTGTCTTACAATCTACGGCGGGTGATGATAGGAATTCAAAGGATGTGTTTTATGACAAGTTTCGAACCGAGCCGAATCGATACAAAGCCCTATTTCGATCGTCGTTTTATA
                                                                                                                     960

N  E  S  R  R  M  N  E  E  K  I  L  E  A  V  K  Y  S  Q  N  L  G  S  H  D  R  R  L  P  L  F  E  K  M  L  K  M  V  Y  T
ACGAATCAAGACGAATGAATGAGGAGAAGATCTTAGAGGCGGTGAAGTATTCGCAAAATTTAGGCTCGCACGATCGTAGGCTACCTCTTTTTGAAAAAATGTTAAAGATGGTTTATACTA
                                                                                                                    1080

T  P  F  Y  P  H  K  S  S  N  M  I  L  A  S  F  L  L  S  I  Q  T  I  T  G  Y  G  R  A  W  V  K  N  V  S  T  E  F  D  K
CACCATTCTACCCGCATAAGAGCTCGAACATGATATTAGCATCTTTCCTATTAAGCATTCAAACCATTACTGGATATGGCAGGGCGTGGGTGAAGAACGTGAGCACCGAGTTCGATAAAC
                                                                                                                    1200

Q  L  K  P  N  P  S  N  L  V  R  D  V  S  D  L  T  R  E  F  F  K  Q  A  Y  V  E  A  K  E  R  R  E  E  M  V  K  P  E  D
AGCTGAAACCGAACCCAAGCAATCTGGTTCGAGATGTTTCGGATTTAACGCGGGAATTCTTCAAGCAAGCATATGTTGAAGCAAAAGAGCGTAGAGAGGAGATGGTTAAACCCGAGGATT
                                                                                                                    1320

L  Y  T  S  M  L  R  L  A  R  N  T  S  S  G  F  S  T  E  I  Y  V  K  K  R  F  G  P  R  L  R  D  K  D  L  V  K  I  N  S
TATACACATCGATGTTACGATTAGCGAGAAATACAAGTTCGGGTTTTTCAACCGAGATTTACGTAAAAAAGAGATTTGGTCCAAGATTAAGGGATAAGGATTTGGTTAAGATCAATTCAA
                                                                                                                    1440

R  I  K  A  L  V  I  F  T  K  G  H  T  V  F  T  D  E  E  L  H  K  K  Y  N  S  V  E  L  Y  Q  T  K  G  S  R  D  V  P  I
GGATTAAGGCTTTGGTTATTTTCACTAAGGGACATACTGTATTCACTGATGAAGAACTTCATAAAAAATATAATAGCGTTGAACTATATCAAACTAAAGGCTCAAGAGATGTACCGATCA
                                                                                                                    1560

K  A  T  R  T  I  Y  S  I  N  L  S  V  L  V  P  Q  L  I  V  T  L  P  L  N  E  Y  F  S  R  V  G  G  I  T  R  P  D  Y  K
AGGCCACAAGAACGATATATTCAATCAATCTCTCAGTGTTGGTACCGCAGTTGATTGTTACTTTACCTTTAAATGAATATTTTTCTAGGGTTGGCGGGATAACTCGTCCGGATTACAAAA
                                                                                                                    1680
```

```
K  I  G  G  K  V  I  V  G  D  L  E  A  T  G  S  R  V  M  D  A  A  D  C  F  R  N  S  A  D  R  D  I  F  T  I  A  I  D  Y
AGATCGGAGGAAAGGTGATTGTCGGAGATTTAGAAGCTACGGGGTCGCGCGTGATGGACGCAGCTGATTGCTTCCGTAACTCTGCCGATCGCGACATATTCACAATCGCAATTGACTATA
                                                                                                                    1800

S  E  Y  D  T  H  L  T  R  H  N  F  R  T  G  M  L  Q  G  I  R  E  A  M  A  P  Y  R  A  L  R  Y  E  G  Y  T  L  E  Q  I
GTGAATACGATACACACCTAACGCGGCATAATTTCCGAACCGGCATGCTCCAAGGGATCAGAGAGGCTATGGCTCCCTATCGGGCTTTGCGATATGAAGGTTATACGTTAGAACAAATCA
                                                                                                                    1920
I  D  F  G  Y  G  E  G  R  V  A  N  T  L  W  N  G  K  R  R  L  F  K  T  T  F  D  A  Y  I  R  L  D  E  S  E  R  D  K  G
TAGATTTTGGATATGGAGAGGGGAGGGTAGCGAATACGTTGTGGAACGGAAAGCGAAGACTGTTTAAGACTACATTTGACGCGTATATACGATTAGATGAGAGCGAGCGAGACAAAGGTA
                                                                                                                    2040

S  F  K  V  P  K  G  V  L  P  V  S  S  V  D  V  A  N  R  I  A  V  D  K  G  F  D  T  L  I  A  A  T  D  G  S  D  L  A  L
GTTTCAAGGTCCCCAAGGGAGTGCTTCCAGTATCGAGTGTTGACGTTGCGAATCGAATCGCGGTGGACAAAGGATTCGACACGCTTATCGCGGCAACGGATGGAAGCGATTTGGCTTTGA
                                                                                                                    2160

I  D  T  H  L  S  G  E  N  S  T  L  I  A  N  S  M  H  N  M  A  I  G  T  L  I  Q  R  A  V  G  R  E  Q  P  G  I  L  T  F
TTGATACACACCTTTCCGGCGAGAATTCGACTCTTATCGCCAATTCGATGCACAATATGGCTATTGGAACCTTGATACAACGGGCCGTTGGGAGGGAGCAGCCAGGGATTCTTACCTTCT
                                                                                                                    2280

L  S  E  Q  Y  V  G  D  D  T  L  F  Y  T  K  L  H  T  T  D  I  T  V  F  D  K  V  A  A  S  I  F  D  T  V  A  K  C  G  H
TATCGGAACAATACGTGGGGGACGATACACTGTTTTACACAAAACTACATACTACAGATATTACGGTTTTCGATAAGGTGGCGGCTTCAATTTTTGATACCGTGGCGAAGTGTGGACATG
                                                                                                                    2400

E  A  S  P  S  K  T  M  M  T  P  Y  S  V  E  K  T  Q  T  H  A  K  Q  G  C  Y  V  P  Q  D  R  M  M  I  I  S  S  E  R  R
AAGCTTCACCTAGTAAAACGATGATGACGCCATACTCTGTGGAAAAAACGCAAACGCATGCAAAACAGGGTTGTTACGTACCACAGGATCGTATGATGATTATCTCATCAGAAAGGAGGA
                                                                                                                    2520

K  D  I  E  D  V  Q  G  Y  V  R  S  Q  V  Q  T  M  I  T  K  V  S  R  G  F  C  H  D  L  A  Q  L  I  L  M  L  K  T  T  F
AGGATATCGAAGATGTGCAGGGATACGTGCGTTCGCAAGTGCAAACGATGATAACGAAAGTGAGTAGAGGATTTTGTCACGATTTAGCGCAGCTAATATTGATGCTTAAGACTACCTTTA
                                                                                                                    2640

I  G  A  W  K  M  K  R  T  I  K  E  N  A  M  Y  R  D  R  K  F  D  S  N  D  E  D  G  F  T  L  I  Q  I  R  N  P  L  A  L
TTGGAGCGTGGAAGATGAAGCGGACTATTAAGGAAAATGCGATGTATCGCGACAGAAAGTTTGATTCGAACGATGAGGATGGGTTTACGCTAATACAGATCCGGAATCCCTTAGCGTTAT
                                                                                                                    2760

Y  V  P  I  G  W  N  G  Y  G  A  H  P  A  A  L  N  I  V  M  T  E  E  M  Y  V  D  S  I  M  I  S  K  L  D  E  I  M  A  P
ATGTTCCTATAGGTTGGAATGGGTACGGTGCACATCCAGCAGCTCTTAATATCGTTATGACGGAAGAGATGTATGTAGATTCGATCATGATCTCAAAGCTGGATGAGATTATGGCGCCGA
                                                                                                                    2880

I  R  R  I  V  H  D  I  P  P  C  W  N  E  T  Q  G  D  K  R  G  L  I  S  A  T  K  L  S  F  F  S  K  M  A  R  P  A  V  Q
TAAGGAGGATTGTGCATGATATTCCGCCATGTTGGAACGAGACTCAGGGAGACAAGCGCGGACTGATCAGTGCAACCAAACTGAGTTTCTTTTCGAAGATGGCTAGGCCAGCTGTCCAAG
                                                                                                                    3000

A  A  L  S  D  P  Q  I  M  N  L  V  E  E  L  P  L  G  E  F  S  P  G  R  I  S  R  T  M  M  H  S  A  L  L  K  E  S  S  A
CCGCTTTAAGCGATCCGCAAATAATGAATCTGGTCGAAGAACTACCGCTTGGAGAGTTTTCACCTGGACGCATTTCAAGAACTATGATGCATAGTGCTCTTCTGAAGGAGTCTAGCGCTA
                                                                                                                    3120

K  A  L  L  S  S  G  Y  R  L  E  Y  Q  K  A  L  N  G  W  I  A  Q  V  S  M  R  L  G  E  E  S  G  V  I  S  T  S  Y  A  K
AGGCGTTATTATCTAGTGGTTATAGACTAGAATATCAGAAGGCTTTGAACGGTTGGATTGCGCAAGTTTCAATGCGTCTGGGAGAGGAGTCTGGAGTAATATCGACATCCTATGCGAAAC
                                                                                                                    3240

L  F  D  V  Y  F  E  G  E  L  D  G  A  P  Y  M  F  P  D  Q  N  L  S  P  Q  F  Y  I  Q  K  M  M  I  G  P  R  V  S  S  R
TCTTCGATGTGTACTTCGAAGGTGAGTTGGACGGAGCACCCTATATGTTTCCAGACCAAAATTTGTCTCCGCAGTTCTATATACAGAAGATGATGATTGGCCCACGAGTTAGCTCACGAG
                                                                                                                    3360
```

L1 (*Continued*)

```
V  R  N  S  Y  V  D  R  I  D  V  I  L  R  K  D  V  V  M  R  G  F  I  T  A  N  T  I  L  N  V  I  E  K  L  G  T  N  H  S
TGCGGAATTCTTATGTTGATCGAATTGATGTGATATTAAGAAAGGATGTCGTAATGCGAGGTTTTATTACTGCCAATACGATTCTGAACGTAATTGAAAAATTAGGGACTAATCACTCAG
                                                                                                                    3480

V  G  D  L  V  T  V  F  T  L  M  N  I  E  T  R  V  A  E  E  L  A  E  Y  M  T  S  E  K  I  R  F  D  A  L  K  L  L  K  K
TGGGAGATCTGGTTACGGTCTTCACGCTTATGAATATCGAAACACGTGTGGCTGAAGAGTTAGCTGAATATATGACTTCAGAGAAGATACGATTCGATGCGTTAAAACTTCTAAAGAAAG
                                                                                                                    3600

G  I  A  G  D  E  F  T  M  S  L  N  V  A  T  Q  D  F  I  D  T  Y  L  A  Y  P  Y  Q  L  T  K  T  E  V  D  A  I  S  L  Y
GGATCGCTGGCGACGAATTCACCATGTCGTTGAATGTCGCTACCCAGGACTTTATTGACACCTACCTCGCCTATCCCTATCAGTTGACGAAAACGGAAGTTGATGCCATATCGTTGTATT
                                                                                                                    3720

C  T  Q  M  V  M  L  R  A  A  L  G  L  P  K  K  K  M  K  I  V  V  T  D  D  A  K  K  R  Y  K  I  R  L  Q  R  F  R  T  H
GCACGCAGATGGTTATGCTACGTGCGGCACTCGGGTTACCAAAAAAGAAGATGAAAATTGTTGTAACTGATGATGCGAAGAAAAGATACAAGATACGTTTGCAGAGGTTTAGAACGCACG
                                                                                                                    3840

V  P  K  I  K  V  L  K  K  L  I  D  P  N  R  M  T  V  R  N  L  E  N  Q  F  V  *
TACCTAAGATTAAAGTTCTTAAGAAGTTGATCGATCCAAATAGAATGACAGTTAGAAATCTTGAGAACCAATTCGTTTGAGAGCACGCGCGAGCACGCGCCGCATTACACTTAC
                                                                                                                3954
```

L2

```
                   M  E  E  F  V  I  P  V  F  S  E  R  D  I  P  Y  S  L  L  N  H  Y  P  L  A  I  Q  I  D  V  K  V  D  D
GTTAAAAGAGTGTTCTACCATGGAGGAATTCGTCATACCAGTGTTTTCTGAGAGAGATATTCCATATTCACTGCTAAATCATTACCCACTAGCTATACAGATAGATGTTAAAGTTGATGA
                                                                                                                     120

  E  G  G  K  H  N  L  I  K  I  P  E  S  D  M  I  D  V  P  R  L  S  I  I  E  A  L  N  Y  R  P  K  R  N  D  G  V  V  V  P
TGAAGGTGGGAAACATAATCTTATCAAGATACCTGAATCGGATATGATCGACGTACCGAGGTTAAGTATCATAGAGGCTCTAAATTATAGACCGAAAAGGAACGATGGTGTGGTTGTGCC
                                                                                                                     240

  R  L  L  D  I  T  L  R  A  Y  D  N  R  K  S  A  K  N  A  K  G  V  E  F  M  T  D  T  K  W  M  K  W  A  I  D  D  K  M  D
GAGATTGCTAGATATAACGCTACGCGCTTATGATAATAGGAAATCCGCAAAGAACGCGAAAGGAGTTGAATTTATGACAGATACTAAGTGGATGAAATGGGCAATTGACGACAAAATGGA
                                                                                                                     360

  I  Q  P  L  K  V  T  L  D  N  H  C  S  V  N  H  Q  L  F  N  C  I  V  K  A  R  S  A  N  A  D  T  I  Y  Y  D  Y  Y  P  L
TATACAACCCCTTAAGGTTACTCTGGATAACCACTGTTCCGTTAACCATCAACTCTTCAATTGTATCGTTAAGGCGAGGTCAGCAAATGCAGACACTATTTATTACGACTATTACCCATT
                                                                                                                     480

  E  N  G  A  K  R  C  N  H  T  N  L  D  L  L  R  S  L  T  T  T  E  M  F  H  I  L  Q  G  A  A  Y  A  L  K  T  T  Y  E  L
GGAAAATGGAGCGAAAAGATGTAATCATACGAACCTGGATTTATTACGAAGTTTAACAACGACGGAGATGTTCCACATCTTACAGGGAGCGGCTTACGCCTTAAAAACAACGTATGAATT
                                                                                                                     600

  V  A  H  S  E  R  E  N  M  S  E  S  Y  Q  V  G  T  Q  R  W  I  Q  L  R  K  G  T  K  I  G  Y  R  G  Q  P  Y  E  R  F  I
AGTAGCTCATTCCGAAAGAGAAAACATGAGCGAGAGTTATCAGGTTGGGACGCAGAGATGGATACAGCTGCGTAAAGGTACTAAAATTGGATATAGAGGGCAGCCGTACGAACGATTTAT
                                                                                                                     720

  S  S  L  V  Q  V  I  I  K  G  K  I  P  D  E  I  R  T  E  I  A  E  L  N  R  I  K  D  E  W  K  N  A  A  Y  D  R  T  E  I
ATCCAGCTTAGTTCAAGTAATCATTAAAGGGAAAATTCCGGATGAGATTAGAACCGAAATCGCTGAACTTAATAGAATAAAAGATGAGTGGAAAAACGCCGCGTATGACAGAACTGAGAT
                                                                                                                     840

  R  A  L  E  L  C  K  I  L  S  A  I  G  R  K  M  L  D  V  Q  E  E  P  K  D  E  M  A  L  S  T  R  F  Q  F  K  L  D  E  K
ACGTGCATTGGAGCTCTGCAAGATACTTTCGGCGATCGGACGTAAAATGTTAGATGTTCAAGAAGAGCCAAAGGATGAGATGGCGTTATCTACTAGATTTCAGTTCAAGTTAGACGAGAA
                                                                                                                     960

  F  I  R  T  D  Q  E  H  V  N  I  F  K  V  G  G  S  A  T  D  D  G  R  F  Y  A  L  I  A  I  A  G  T  D  T  Q  Q  G  R  V
ATTCATTAGGACTGATCAAGAACACGTAAATATATTTAAAGTAGGTGGTTCCGCAACAGATGATGGGAGATTTTACGCGCTAATCGCTATAGCAGGTACCGATACGCAGCAAGGCCGTGT
                                                                                                                    1080

  W  R  T  N  P  Y  P  C  L  R  G  A  L  I  A  A  E  C  E  L  G  D  V  Y  F  T  L  R  Q  T  Y  K  W  S  L  R  P  E  Y  G
TTGGAGAACGAACCCTTATCCATGTTTGCGTGGTGCGCTAATTGCTGCTGAGTGTGAGCTCGGCGATGTGTACTTCACATTGCGTCAAACGTATAAATGGAGTCTGCGTCCAGAATATGG
                                                                                                                    1200

  Q  R  E  R  P  L  E  D  N  K  Y  V  F  A  R  L  N  L  F  D  T  N  L  A  V  G  D  E  I  I  H  W  R  Y  E  V  Y  Q  P  K
TCAGCGTGAGAGGCCGTTAGAAGACAATAAATATGTGTTCGCTCGTCTTAATCTATTTGATACTAACTTAGCCGTTGGGGATGAAATCATTCACTGGCGTTATGAGGTTTATCAGCCTAA
                                                                                                                    1320

  E  T  T  H  D  D  G  Y  I  C  V  S  Q  K  G  D  D  E  L  L  C  E  V  D  E  D  R  Y  K  E  M  F  D  R  M  I  Q  G  G  W
AGAAACAACTCATGATGATGGATATATCTGTGTGTCACAGAAAGGTGATGATGAATTATTATGTGAGGTTGATGAAGATAGGTATAAGGAGATGTTTGATCGTATGATTCAAGGCGGATG
                                                                                                                    1440

  D  Q  E  R  F  K  L  H  N  I  L  T  E  P  N  L  L  T  I  D  F  E  K  D  A  Y  L  G  A  R  S  E  L  V  F  P  P  Y  Y  D
GGATCAGGAACGGTTCAAATTACACAACATATTGACTGAACCAAACCTACTTACCATCGATTTCGAGAAGGATGCTTACCTTGGCGCGCGTTCTGAATTAGTATTCCCGCCCTATTATGA
                                                                                                                    1560

  K  W  I  N  S  P  M  F  N  A  R  L  K  I  A  R  G  E  I  A  T  W  K  A  D  D  P  W  S  N  R  A  V  H  G  Y  I  K  T  S
CAAATGGATCAATTCACCGATGTTTAATGCCCGGCTTAAAATTGCGCGTGGCGAGATTGCAACCTGGAAGGCGGATGATCCTTGGAGTAATCGGGCGGTTCATGGGTATATAAAGACGTC
                                                                                                                    1680
```

L2 (*Continued*)

```
  A  E  S  L  E  Y  A  L  G  P  Y  Y  D  L  R  L  Q  L  F  G  D  T  L  S  L  G  Q  R  Q  S  A  V  F  E  H  M  A  Q  Q  D
AGCGGAATCTTTAGAGTATGCGTTAGGGCCATATTATGACCTACGTCTGCAGTTATTCGGCGACACCCTCAGTCTGGGACAAAGACAGTCAGCTGTATTCGAACATATGGCTCAACAAGA
                                                                                                                    1800
  D  F  S  T  L  T  D  Y  T  K  G  R  T  V  C  P  H  S  G  G  T  F  Y  T  F  R  K  V  A  L  I  I  L  S  N  Y  E  R  L  D
TGATTTTTCTACGCTAACGGATTATACGAAAGGAAGAACAGTGTGCCCTCACTCAGGCGGACTTTTTACACTTTCCGCAAGGTAGCGTTGATTATTTTATCGAATTACGAACGATTGGA
                                                                                                                    1920
  P  S  L  H  E  G  R  E  H  E  T  Y  M  H  P  A  V  N  D  V  F  R  R  H  V  L  E  M  K  D  F  S  Q  L  I  C  F  V  F  D
CCCAAGTTTACATGAGGGGAGGGAACATGAGACATACATGCACCCGGCAGTTAATGATGTCTTTAGACGACACGTTTTAGAAATGAAAGATTTTTCCCAGCTGATTTGTTTTGTGTTCGA
                                                                                                                    2040
  Y  I  F  E  K  H  V  Q  L  R  N  A  K  E  A  R  R  I  I  Y  L  I  Q  N  T  S  G  A  Y  R  L  D  V  L  R  E  A  F  P  N
CTACATTTTTGAGAAACATGTACAGCTACGTAATGCCAAAGAGGCGAGAAGGATTATATATTTAATACAGAATACGTCCGGTGCGTATCGATTAGATGTTTTACGCGAAGCTTTCCCCAA
                                                                                                                    2160
  F  L  K  H  V  M  N  L  R  D  V  K  R  I  C  D  L  N  V  I  N  F  F  P  L  L  F  L  V  Q  D  N  I  S  Y  W  H  R  Q  W
CTTTTTAAAGCACGTTATGAATTTACGTGATGTGAAACGTATATGCGATCTGAATGTGATTAATTTCTTTCCATTGTTATTTTTGGTTCAAGATAATATTTCATATTGGCATAGACAGTG
                                                                                                                    2280
  S  I  P  M  I  L  F  D  Q  V  I  R  L  I  P  V  E  V  G  A  Y  A  N  R  F  G  L  K  S  F  F  N  F  I  R  F  H  P  G  D
GTCGATCCCAATGATTTTATTTGATCAAGTTATCAGATTAATTCCCGTCGAAGTCGGGGCATACGCAAATAGATTTGGACTTAAGAGCTTCTTCAATTTCATTAGGTTTCACCCCGGCGA
                                                                                                                    2400
  S  K  K  R  Q  D  A  D  D  T  H  K  E  F  G  S  I  C  F  E  Y  Y  T  T  T  K  I  S  Q  G  E  I  D  V  P  V  V  T  S  K
TTCAAAGAAGAGACAGGATGCGGATGACACCCATAAAGAATTCGGCTCAATATGTTTCGAGTATTACACTACGACAAAGATCTCTCAAGGTGAGATAGATGTTCCGGTCGTTACGTCTAA
                                                                                                                    2520
  L  D  T  L  K  L  H  V  A  S  L  C  A  G  L  A  D  S  L  V  Y  T  L  P  V  A  H  P  K  K  S  I  V  L  I  I  V  G  D  D
ATTAGATACACTTAAGTTGCACGTAGCTTCCTTATGCGCGGGGCTAGCTGACTCTCTAGTGTATACGCTGCCTGTAGCGCACCCGAAGAAAAGTATCGTTCTAATTATTGTGGGGGATGA
                                                                                                                    2640
  K  L  E  P  Q  I  R  S  E  Q  I  V  N  K  Y  Y  Y  S  R  R  H  I  S  G  V  V  S  I  C  V  N  Q  G  G  Q  L  K  V  H  S
TAAGCTTGAACCACAAATACGCTCAGAGCAAATCGTAAATAAGTACTATTATTCACGGAGGCACATCTCAGGGGTTGTTTCGATCTGCGTGAACCAGGGTGGACAACTGAAGGTGCATTC
                                                                                                                    2760
  M  G  I  T  R  H  R  I  C  D  K  S  I  L  K  Y  K  C  K  V  V  L  V  R  M  P  G  H  V  F  G  N  D  E  L  M  T  K  L  L
CATGGGGATAACGAGACATCGGATATGTGATAAGTCGATATTGAAATATAAATGCAAAGTAGTATTAGTTAGGATGCCAGGACATGTCTTCGGAAATGATGAATTAATGACGAAGCTACT
                                                                                                                    2880
  N  V  *
AAATGTTTAGGTCCTGTGACATGGACCGGTAGCCTCTTACACTTAC
                                          2926
```

L3

```
                   M  A  A  Q  N  E  Q  R  P  E  R  I  K  T  T  P  Y  L  E  G  D  V  L  S  S  D  S  G  P  L  L  S  V  F
GTTAAATTTCCGTAGCCATGGCTGCTCAGAATGAGCAACGTCCGGAACGAATAAAAACGACGCCGTATTTAGAAGGAGATGTGCTTTCGAGTGACTCAGGGCCACTGCTTTCTGTGTTCG
                                                                                                                     120

A  L  Q  E  I  M  Q  K  V  R  Q  V  Q  A  D  Y  M  T  A  T  R  E  V  D  F  T  V  P  D  V  Q  K  I  L  D  D  I  K  T  L
CGTTACAAGAGATAATGCAAAAGGTGAGGCAAGTGCAAGCTGACTATATGACGGCAACACGAGAGGTTGATTTTACAGTACCGGATGTACAAAAGATTCTTGATGACATTAAAACGTTAG
                                                                                                                     240

A  A  E  Q  V  Y  K  I  V  K  V  P  S  I  S  F  R  H  I  V  M  Q  S  R  D  R  V  L  R  V  D  T  Y  Y  E  E  M  S  Q  V
CTGCAGAACAAGTGTACAAAATCGTTAAAGTTCCTAGTATTTCATTCAGACATATCGTAATGCAGTCAAGAGATCGAGTTTTACGAGTGGATACCTACTATGAAGAGATGTCACAGGTTG
       250       260       270       280       290       300       310       320       330       340       350       360

G  D  V  I  T  E  D  E  P  E  K  F  Y  S  T  I  I  K  K  V  R  F  I  R  G  K  G  S  F  I  L  H  D  I  P  T  R  D  H  R
GAGATGTTATAACGGAAGATGAACCAGAAAAATTCTATTCAACTATAATCAAGAAAGTGCGGTTCATACGCGGAAAAGGATCCTTTATATTACATGATATTCCGACCAGAGATCACCGCG
                                                                                                                     480

G  M  E  V  A  E  P  E  V  L  G  V  E  F  K  N  V  L  P  V  L  T  A  E  H  R  A  M  I  Q  N  A  L  D  G  S  I  I  E  N
GTATGGAGGTTGCTGAACCAGAGGTGTTAGGAGTTGAATTCAAGAATGTACTACCTGTGTTGACTGCCGAGCATCGCGCAATGATTCAGAACGCATTGGATGGATCGATAATCGAGAACG
                                                                                                                     600

G  N  V  A  T  R  D  V  D  V  F  I  G  A  C  S  E  P  I  Y  R  I  Y  N  R  L  Q  G  Y  I  E  A  V  Q  L  Q  E  L  R  N
GAAACGTAGCTACACGAGACGTCGACGTATTCATAGGCGCTTGTTCGGAACCAATCTATCGTATATATAATAGATTGCAAGGGTATATTGAAGCGGTGCAGTTACAAGAGTTAAGGAACT
                                                                                                                     720

S  I  G  W  L  E  R  L  G  Q  R  K  R  I  T  Y  S  Q  E  V  L  T  D  F  R  R  Q  D  T  I  W  V  L  A  L  Q  L  P  V  N
CAATTGGGTGGTTAGAAAGGTTAGGGCAGAGGAAAAGAATCACGTATTCGCAAGAGGTTCTGACTGATTTTAGGAGGCAGGATACAATTTGGGTTTTAGCCTTACAGCTACCGGTTAATC
                                                                                                                     840

P  Q  V  V  W  D  V  P  R  S  S  I  A  N  L  I  M  N  I  A  T  C  L  P  T  G  E  Y  I  A  P  N  P  R  I  S  S  I  T  L
CGCAGGTAGTGTGGGATGTGCCGCGCAGCTCTATCGCCAACTTAATCATGAATATAGCAACGTGCTTACCCACGGGGGAATACATCGCGCCAAACCCAAGAATTTCATCGATTACGCTGA
                                                                                                                     960

T  Q  R  I  T  T  T  G  P  F  A  I  L  T  G  S  T  P  T  A  Q  Q  L  N  D  V  R  K  I  Y  L  A  L  M  F  P  G  Q  I  V
CCCAGAGAATAACGACAACTGGGCCATTTGCTATTCTAACAGGATCAACTCCAACCGCACAGCAACTTAATGATGTTAGGAAGATCTATTTAGCGCTAATGTTTCCTGGACAGATTGTAC
                                                                                                                    1080

L  D  L  K  I  D  P  G  E  R  M  D  P  A  V  R  M  V  A  G  V  V  G  H  L  L  F  T  A  G  G  R  F  T  N  L  T  Q  N  M
TTGATCTAAAAATTGATCCTGGTGAGAGGATGGATCCGGCAGTAAGAATGGTCGCTGGCGTTGTAGGTCATTTGCTCTTTACAGCAGGTGGAAGATTCACGAATTTAACACAAAATATGG
                                                                                                                    1200

A  R  Q  L  D  I  A  L  N  D  Y  L  L  Y  M  Y  N  T  R  V  Q  V  N  Y  G  P  T  G  E  P  L  D  F  Q  I  G  R  N  Q  Y
CGAGACAGCTTGATATAGCCTTAAACGATTATTTACTTTATATGTATAACACCAGAGTTCAAGTCAATTATGGTCCAACAGGTGAGCCGTTAGATTTCCAGATTGGAAGAAATCAGTATG
                                                                                                                    1320

D  C  N  V  F  R  A  D  F  A  T  G  T  G  Y  N  G  W  A  T  I  D  V  E  Y  R  D  P  A  P  Y  V  H  A  Q  R  Y  I  R  Y
ACTGTAATGTTTTCAGAGCAGATTTTGCGACAGGAACGGGATACAATGGATGGGCTACGATAGATGTTGAGTATAGAGATCCGGCCCCTTACGTGCATGCGCAGCGCTACATACGTTATT
                                                                                                                    1440

C  G  I  D  S  R  E  L  I  N  P  T  T  Y  G  I  G  M  T  Y  H  C  Y  N  E  M  L  R  M  L  V  A  A  G  R  D  S  E  A  A
GTGGTATCGATTCGCGTGAGTTAATTAATCCGACAACATATGGCATTGGGATGACTTATCATTGCTACAATGAGATGTTACGAATGCTGGTTGCTGCAGGGAGAGACTCTGAGGCGGCGT
                                                                                                                    1560

Y  F  R  S  M  L  P  F  H  M  V  R  F  A  R  I  N  Q  I  I  N  E  D  L  H  S  V  F  S  L  P  D  D  M  F  N  A  L  L  P
ACTTTCGCAGCATGCTACCCTTTCACATGGTGAGGTTTGCTAGGATAAACCAAATCATAAACGAAGATTTACACTCCGTGTTTTCGTTGCCAGATGATATGTTCAACGCATTATTACCCG
                                                                                                                    1680
```

L3 (*Continued*)

```
D  L  I  A  G  A  H  Q  N  A  D  P  V  V  L  D  V  S  W  I  S  L  W  F  A  F  N  R  S  F  E  P  T  H  R  N  E  M  L  E
ACCTAATTGCTGGGGCGCATCAGAACGCCGACCCAGTTGTGTTAGATGTGAGTTGGATATCACTGTGGTTTGCTTTCAACAGATCGTTTGAACCAACGCATAGGAATGAGATGCTTGAAA
                                                                                                                    1800

I  A  P  L  I  E  S  V  Y  A  S  E  L  S  V  M  K  V  D  M  R  H  L  S  L  M  Q  R  R  F  P  D  V  L  I  Q  A  R  P  S
TCGCCCCACTGATCGAGTCCGTTTATGCGTCGGAACTTTCAGTGATGAAGGTCGATATGCGACACTTGTCATTGATGCAGAGAAGATTTCCAGATGTTTTAATCCAAGCGAGGCCGTCCC
                                                                                                                    1920

H  F  W  K  A  V  L  N  D  S  P  E  A  V  K  A  V  M  N  L  S  H  S  H  N  F  I  N  I  R  D  M  M  R  W  V  L  L  P  S
ATTTCTGGAAAGCGGTTCTGAATGATAGCCCGGAGGCGGTGAAAGCAGTCATGAACTTGTCGCATTCGCATAATTTCATCAATATAAGGGATATGATGCGTTGGGTACTGCTCCCATCAC
                                                                                                                    2040

L  Q  P  S  L  K  L  V  L  E  E  E  A  W  A  A  A  N  D  F  E  D  L  M  L  T  D  Q  V  Y  M  H  R  D  M  L  P  E  P  R
TACAACCATCGTTAAAGCTCGTATTAGAAGAGGAGGCATGGGCCGCTGCAAACGATTTCGAGGATCTGATGCTTACTGATCAAGTTTATATGCATCGAGATATGTTGCCAGAACCGCGGT
                                                                                                                    2160

L  D  D  I  E  R  F  R  Q  E  G  F  Y  Y  T  N  M  L  E  A  P  P  E  I  D  R  V  V  Q  Y  T  Y  E  I  A  R  L  Q  A  N
TGGATGATATTGAGAGGTTCAGACAGGAAGGTTTCTATTACACGAATATGTTAGAAGCCCCACCAGAAATAGATCGTGTAGTTCAGTACACTTATGAGATTGCACGTTTACAAGCGAACA
                                                                                                                    2280

M  G  Q  F  R  A  A  L  R  R  I  M  D  D  D  D  W  V  R  F  G  G  V  L  R  T  V  R  V  K  F  F  D  A  R  P  P  D  D  I
TGGGCCAGTTTCGAGCGGCTTTAAGGCGCATTATGGATGATGACGACTGGGTGAGGTTTGGCGGTGTTTTACGCACTGTACGCGTTAAATTCTTTGATGCGCGACCCCCAGACGACATTC
                                                                                                                    2400

L  Q  G  L  P  F  S  Y  D  T  N  E  K  G  G  L  S  Y  A  T  I  K  Y  A  T  E  T  T  I  F  Y  L  I  Y  N  V  E  F  S  N
TACAAGGCTTACCTTTCAGCTATGATACAAATGAGAAAGGTGGATTATCATATGCGACAATTAAGTATGCTACTGAGACCACAATTTTCTATCTGATATATAATGTCGAATTCTCGAACA
                                                                                                                    2520

T  P  D  S  L  V  L  I  N  P  T  Y  T  M  T  K  V  F  I  N  K  R  I  V  E  R  V  R  V  G  Q  I  L  A  V  L  N  R  R  F
CGCCTGATTCTTTGGTGTTGATTAATCCAACATACACGATGACTAAAGTTTTCATTAACAAGAGAATTGTTGAGCGGGTACGGGTCGGACAGATTCTGGCCGTATTGAACAGAAGATTTG
                                                                                                                    2640

V  A  Y  K  G  K  M  R  I  M  D  I  T  Q  S  L  K  M  G  T  K  L  A  A  P  T  V  *
TGGCATACAAAGGAAAAATGAGAATCATGGACATCACCCAATCGCTCAAGATGGGCACCAAGCTGGCTGCGCCAACTGTGTAGATGTGCGACCGATCTATGCACTTGGTAGCGGCAGCGG
                                                                                                                    2760

AAACACACTTAC
        2772
```

M4

```
        M  P  E  P  H  A  V  L  Y  V  T  N  E  L  S  H  I  V  K  N  E  F  L  P  I  W  K  L  T  G  D  E  S  L  N  D  L
GTTAAAACATGCCTGAGCCACACGCAGTACTCTATGTCACAAATGAGCTCTCTCATATCGTTAAGAATGAATTTTTGCCAATATGGAAACTAACTGGAGATGAGAGTTTAAATGACTTAT
                                                                                                                     120

W  L  E  N  G  K  Y  A  T  D  V  Y  A  Y  G  D  V  S  K  W  T  I  R  Q  L  R  G  H  G  F  I  F  I  S  T  H  K  N  V  Q
GGCTTGAAAATGGAAAGTACGCGACAGACGTATATGCGTACGGAGACGTTTCAAAGTGGACGATACGACAATTACGGGGTCATGGATTCATTTTTATAAGCACGCACAAGAACGTACAGC
                                                                                                                     240

L  A  D  I  I  K  T  V  D  V  R  V  P  R  E  V  A  R  S  H  D  M  K  A  F  E  N  E  I  G  R  R  R  I  R  M  R  K  G  F
TAGCTGACATTATTAAGACGGTTGACGTCCGCGTTCCTCGGGAGGTTGCAAGAAGTCATGATATGAAAGCTTTTGAAAATGAGATCGGCAGACGCAGGATCCGTATGCGCAAGGGATTTG
                                                                                                                     360

G  D  A  L  R  N  Y  A  F  K  M  A  I  E  F  H  G  S  E  A  E  T  L  N  D  A  N  P  R  L  H  K  I  Y  G  M  P  E  I  P
GTGATGCATTGAGAAATTACGCCTTTAAGATGGCTATCGAATTTCACGGGTCGGAGGCGGAAACGTTGAACGATGCAAATCCTCGGTTACATAAAATTTATGGAATGCCGGAAATACCGC
                                                                                                                     480

P  L  Y  M  E  Y  A  E  I  G  T  R  F  D  D  E  P  T  D  E  K  L  V  S  M  L  H  Y  I  V  Y  S  A  E  E  V  H  Y  I  G
CATTATACATGGAATATGCGGAAATAGGGACTAGATTTGACGATGAGCCGACTGATGAAAAGTTAGTATCAATGCTTCATTATATCGTTTACAGTGCCGAAGAGGTGCACTATATTGGAT
                                                                                                                     600

C  G  D  L  R  T  L  M  Q  F  K  K  R  S  P  G  R  F  R  R  V  L  W  H  V  Y  D  P  I  A  P  E  C  S  D  P  N  V  I  V
GTGGTGACCTACGTACCCTAATGCAGTTCAAGAAACGATCACCAGGACGGTTTAGAAGGGTGTTATGGCACGTATATGACCCAATAGCACCTGAGTGTTCAGATCCAAACGTTATAGTTC
                                                                                                                     720

H  N  I  M  V  D  S  K  K  D  I  L  K  H  M  N  F  L  K  R  V  E  R  P  F  I  W  D  V  S  S  D  R  S  Q  M  N  D  H  E
ATAATATTATGGTGGATTCAAAGAAAGACATTTTGAAACATATGAATTTTTTGAAACGTGTTGAGAGACCTTTCATATGGGATGTATCGTCAGATCGATCGCAGATGAATGATCATGAGT
                                                                                                                     840

W  E  T  T  R  F  A  E  D  R  L  G  E  E  I  A  Y  E  M  G  G  A  F  S  S  A  L  I  K  H  R  I  P  N  S  K  D  E  Y  H
GGGAGACGACGAGGTTTGCGGAGGATAGATTGGGTGAGGAAATAGCTTATGAAATGGGTGGTGCATTTTCCAGTGCATTGATCAAGCACAGGATACCGAATTCAAAAGACGAATATCACT
                                                                                                                     960

C  I  S  T  Y  L  F  P  Q  P  G  A  D  A  D  M  Y  E  L  R  N  F  M  R  L  R  G  Y  S  H  V  D  R  H  M  H  P  D  A  S
GCATTTCGACCTACTTATTCCCCCAACCGGGGGCCGATGCGGATATGTATGAATTAAGGAATTTCATGAGATTGAGAGGTTACTCACACGTGGATCGCCACATGCATCCAGACGCTTCCG
                                                                                                                    1080

V  T  K  V  V  S  R  D  V  R  K  M  V  E  L  Y  H  G  R  D  C  G  S  F  L  K  K  R  L  F  E  H  L  H  I  V  R  K  N  G
TGACGAAAGTTGTTTCACGTGATGTGCGCAAAATGGTCGAATTGTATCACGGTCGCGATTGTGGTAGCTTTCTAAAAAAGAGACTATTTGAACACCTTCATATTGTACGTAAGAATGGAT
                                                                                                                    1200

L  L  H  E  S  D  E  P  R  A  D  L  F  Y  L  T  N  R  C  N  M  G  L  E  P  S  I  Y  E  V  M  K  K  S  V  I  A  T  A  W
TGTTACACGAAAGTGATGAGCCACGAGCGGATCTGTTTTATTTGACCAATCGGTGCAATATGGGATTGGAGCCTAGTATTTATGAGGTAATGAAGAAGTCGGTGATAGCTACTGCTTGGG
                                                                                                                    1320

V  G  R  A  P  L  Y  D  Y  D  D  F  A  L  P  R  S  T  V  M  L  N  G  S  Y  R  D  I  R  I  L  D  G  N  G  A  I  L  F  L
TGGGCCGTGCGCCTTTATATGATTATGATGATTTCGCGTTACCCAGATCTACCGTTATGCTCAACGGATCCTACCGCGATATCAGAATTCTAGATGGCAATGGCGCAATCCTATTCCTAA
                                                                                                                    1440

M  W  R  Y  P  D  I  V  K  K  D  L  T  Y  D  P  A  W  A  M  N  F  A  V  S  L  K  E  P  I  P  D  P  P  V  P  D  I  S  L
TGTGGAGGTACCCGGATATCGTAAAGAAAGATTTGACGTATGACCCTGCGTGGGCGATGAATTTTGCTGTTTCGCTAAAGGAGCCGATACCTGATCCTCCTGTGCCTGATATTTCTTTGT
                                                                                                                    1560

C  R  F  I  G  L  R  V  E  S  S  V  L  R  V  R  N  P  T  L  H  E  T  A  D  E  L  P  T  L  H  E  T  A  D  E  L  K  R  M
GTAGGTTCATCGGACTGCGCGTCGAATCATCCGTGCTGAGGGTCCGAAACCCAACATTACATGAGACGGCTGACGAACTGCCAACATTACATGAGACGGCTGACGAACTGAAACGGATGG
                                                                                                                    1680

G  L  D  L  S  G  H  L  Y  V  T  L  M  S  G  A  Y  V  T  D  L  F  W  W  F  K  M  I  L  D  W  S  A  Q  N  R  E  Q  K  L
GATTGGATTTGTCTGGTCATTTATATGTCACATTAATGTCCGGCGCTTATGTCACAGATCTGTTTTGGTGGTTTAAGATGATTCTAGATTGGTCTGCGCAAAACAGGGAGCAAAAACTAC
                                                                                                                    1800

R  D  L  K  R  S  A  A  E  V  I  E  W  K  E  Q  M  A  E  R  P  W  H  V  R  N  D  L  I  A  A  L  R  E  Y  K  R  K  M  G
GCGATCTAAAGAGGTCTGCGGCGGAAGTAATTGAGTGGAAGGAGCAGATGGCCGAGCGTCCATGGCATGTGAGAAATGATCTGATCGCTGCGCTAAGGGAATACAAACGGAAAATGGGGA
                                                                                                                    1920

                             M  R  E  G  A  S  I  D  S  W  L  E  L  L  R  H  L  *
                             TGAGAGAGGGAGCCTCGATTGATTCGTGGCTAGAATTACTGCGTCACTTATAATGCGTGACTGCTAGGTGAGGGGGGCATGTACAACTTAC
                                                                                                                    2011
```

M5

```
                              M  G  K  I  I  K  S  L  S  R  F  G  K  K  V  G  N  A  L  T  S  N  T  A  K  K  I  Y  S  T
GTTAAAAAGTGTTCTCCTACTCGCAGAAGATGGGGAAGATAATCAAATCGCTTAGTCGTTTTGGAAAAAAGGTCGGGAATGCTTTGACATCGAATACAGCGAAAAAGATTTACTCAACGA
                                                                                                                       120

I  G  K  A  A  E  R  F  A  E  S  E  I  G  A  A  T  I  D  G  L  V  Q  G  S  V  H  S  I  I  T  G  E  S  Y  G  E  S  V  K
TTGGAAAAGCAGCGGAACGTTTTGCAGAAAGTGAAATAGGTGCGGCAACGATTGATGGGTTAGTACAAGGAAGCGTTCACTCAATTATAACAGGAGAATCGTATGGGGAATCGGTGAAAC
                                                                                                                       240

Q  A  V  L  L  N  V  L  G  T  G  E  E  L  P  D  P  L  S  P  G  E  R  G  M  Q  T  K  I  K  E  L  E  D  E  Q  R  N  E  L
AGGCGGTCCTATTGAATGTGTTGGGTACAGGCGAAGAATTGCCTGATCCCCTCAGCCCTGGTGAGCGCGGCATGCAGACGAAGATAAAGGAATTGGAAGATGAGCAACGTAACGAACTTG
                                                                                                                       360

V  R  L  K  Y  N  K  E  I  T  E  K  F  G  K  E  L  G  E  V  Y  D  F  M  N  G  E  A  K  E  V  E  A  V  E  E  Q  Y  T  M
TTAGGCTTAAATATAATAAAGAGATCACCGAGAAATTCGGTAAAGAGCTCGGAGAAGTTTATGACTTCATGAATGGGGAGGCAAAAGAAGTGGAGGCGGTTGAAGAACAGTATACTATGC
                                                                                                                       480

L  C  K  A  V  D  S  Y  E  K  I  L  K  E  E  D  S  K  M  A  I  L  A  R  A  L  Q  R  E  A  A  E  R  S  E  D  E  I  K  M
TATGTAAGGCTGTTGATTCGTATGAGAAAATCCTGAAGGAGGAAGATTCGAAGATGGCTATTTTAGCGCGTGCGTTACAACGAGAGGCAGCGGAAAGAAGTGAAGATGAAATCAAAATGG
                                                                                                                       600

V  K  E  Y  R  Q  K  I  D  A  L  K  S  A  I  E  I  E  R  D  G  M  Q  E  E  A  I  Q  E  I  A  G  M  T  A  D  V  L  E  A
TAAAAGAGTATAGACAGAAGATTGATGCGCTTAAAAGTGCGATAGAGATTGAGCGTGATGGAATGCAGGAAGAGGCTATACAAGAGATCGCTGGAATGACAGCAGACGTGCTAGAGGCGG
                                                                                                                       720

A  S  E  E  V  P  L  I  G  A  G  M  A  T  A  V  A  T  G  R  A  I  E  G  A  Y  K  L  K  K  V  I  N  A  L  S  G  I  D  L
CATCAGAAGAAGTGCCTTTGATAGGTGCCGGTATGGCAACGGCTGTCGCAACGGGACGAGCAATAGAGGGGGCGTATAAATTAAAGAAGGTCATCAATGCACTCAGCGGAATTGACTTAT
                                                                                                                       840

S  H  M  R  S  P  K  I  E  P  T  I  I  A  T  T  L  E  H  R  F  K  D  I  P  D  E  Q  L  A  I  S  V  L  N  K  K  T  A  V
CACATATGCGAAGCCCCAAGATCGAACCTACTATTATCGCCACAACTCTTGAGCATAGGTTTAAAGATATACCAGATGAACAATTGGCTATTAGCGTATTAAATAAAAAGACGGCTGTGG
                                                                                                                       960

A  D  N  C  N  E  I  A  H  I  K  Q  E  I  L  P  K  F  K  Q  I  M  N  E  E  K  E  I  E  G  I  E  D  K  V  I  H  P  R  V
CTGATAATTGCAACGAGATTGCTCATATTAAACAAGAGATACTACCAAAGTTTAAACAGATAATGAATGAAGAGAAGGAGATTGAGGGGATAGAGGATAAAGTCATACACCCGAGAGTAA
                                                                                                                      1080

M  M  R  F  K  I  P  R  T  Q  Q  P  Q  I  H  I  Y  A  A  P  W  D  S  D  D  V  F  F  F  H  C  V  S  Y  H  H  R  N  E  S
TGATGAGGTTTAAGATCCCACGTACGCAGCAGCCCCAAATTCATATCTATGCAGCCCCATGGGATTCGGACGATGTCTTCTTCTTTCACTGTGTGTCTTACCATCATCGGAATGAATCAT
                                                                                                                      1200

F  F  L  G  F  D  L  G  I  D  V  V  H  F  E  D  L  T  S  H  W  H  A  L  G  M  A  Q  E  A  S  G  R  T  L  T  E  A  Y  R
TCTTTCTAGGCTTTGATTTAGGAATTGATGTAGTTCATTTTGAAGATTTAACCAGTCACTGGCACGCGTTAGGGATGGCACAAGAGGCCAGTGGACGGACTTTAACGGAGGCTTATAGAG
                                                                                                                      1320

E  F  L  N  L  S  I  S  S  T  F  S  S  A  I  H  A  R  R  M  I  R  S  R  A  V  H  P  I  F  L  G  S  M  H  Y  D  I  T  Y
AGTTCCTCAATCTTTCGATTTCGAGCACGTTCAGTAGCGCAATACATGCGAGGCGTATGATAAGATCACGCGCTGTACACCCTATCTTTTTAGGTTCAATGCATTACGATATAACTTATG
                                                                                                                      1440

E  A  L  K  N  N  A  Q  R  I  V  Y  D  D  E  L  Q  M  H  I  L  R  G  P  L  H  F  Q  R  R  A  I  L  G  A  L  K  F  G  V
AAGCTCTGAAGAATAATGCGCAGCGGATAGTTTACGATGATGAATTGCAAATGCATATACTTAGAGGACCGTTACACTTCCAGCGCCGAGCAATATTAGGTGCGTTAAAATTCGGAGTTA
                                                                                                                      1560

K  I  L  G  D  K  I  D  V  P  L  F  L  R  N  A  *
AGATATTAGGCGATAAGATAGATGTTCCCCTCTTCTTACGAAATGCTTGAACGCAGCGGGGGAGGACCTTCCACTTAC
                                                                             1638
```

M6

```
                                    M  E  R  F  L  R  K  Y  N  I  S  G  D  Y  A  N  A  T  R  T  F  L  A  I  S  P  Q  W  T
GTTAAAAAAAGTTCTCTAGTTGGCAACCACCAAACATGGAGCGCTTTTTGAGAAAATACAATATCAGTGGGGATTACGCAAATGCCACGAGAACTTTTTTGGCTATTTCACCACAATGGAC
                                                                                                                     120

 C  S  H  L  K  R  N  C  L  F  N  G  M  C  V  K  Q  H  F  E  R  A  M  I  A  A  T  D  A  E  E  P  A  K  A  Y  K  L  V  E
TTGCAGTCACTTAAAAAGGAATTGTCTATTTAATGGAATGTGTGTCAAGCAGCATTTGAGAGAGCGATGATCGCGGCAACTGATGCGGAGGAGCCGGCGAAAGCATACAAATTGGTTGA
                                                                                                                     240

 L  A  K  E  A  M  Y  D  R  E  T  V  W  L  Q  C  F  K  S  F  S  Q  P  Y  E  E  D  V  E  G  K  M  K  R  C  G  A  Q  L  L
ATTGGCAAAGGAAGCAATGTATGATCGGGAAACAGTCTGGCTTCAATGTTTCAAAAGCTTTTCCCAACCATACGAGGAGGATGTCGAAGGGAAGATGAAGCGATGCGGAGCGCAATTGCT
                                                                                                                     360

 E  D  Y  R  K  S  G  M  M  D  E  A  V  K  Q  S  A  L  V  N  S  E  R  I  R  L  D  D  S  L  S  A  M  P  Y  I  Y  V  P  I
TGAGGATTACCGCAAAAGTGGGATGATGGATGAAGCCGTGAAACAATCTGCGCTGGTTAATTCAGAAAGAATTAGATTGGATGATTCACTTTCCGCGATGCCTTACATCTACGTGCCAAT
                                                                                                                     480

 N  D  G  Q  I  V  N  P  T  F  I  S  R  Y  R  Q  I  A  Y  Y  F  Y  N  P  D  A  A  D  D  W  I  D  P  N  L  F  G  I  R  G
CAATGATGGTCAAATTGTTAATCCGACATTTATATCAAGATATCGCCAAATTGCATATTATTTTTACAACCCAGATGCAGCTGATGATTGGATTGATCCAAATCTCTTTGGCATTCGTGG
                                                                                                                     600

 Q  H  N  Q  I  K  R  E  V  E  R  Q  I  N  T  C  P  Y  T  G  Y  R  G  R  V  F  Q  V  M  F  L  P  I  Q  L  I  N  F  L  R
ACAGCACAATCAGATTAAACGTGAGGTTGAGAGACAAATTAACACATGCCCTTACACTGGATACAGAGGTAGAGTGTTTCAAGTAATGTTTTTGCCGATTCAGCTAATCAATTTTTTGAG
                                                                                                                     720

 M  D  D  F  A  K  H  F  N  R  Y  A  S  M  A  I  Q  Q  Y  L  R  V  G  Y  A  E  E  I  R  Y  V  Q  Q  L  F  G  R  V  P  T
AATGGATGATTTTGCGAAGCATTTTAACAGGTACGCCTCGATGGCGATACAACAATATCTGAGAGTTGGTTATGCTGAAGAGATCAGATATGTACAACAGCTCTTCGGAAGGGTCCCAAC
                                                                                                                     840

 G  E  F  P  L  H  Q  M  M  L  M  R  R  D  L  P  T  R  D  R  S  I  V  E  A  R  V  R  R  S  G  D  E  S  W  Q  S  W  L  L
AGGTGAATTTCCATTACACCAGATGATGCTGATGAGACGCGATCTCCCAACGCGCGATCGCAGTATTGTGGAGGCGCGGGTGAGGAGATCAGGTGATGAGAGCTGGCAAAGCTGGCTACT
                                                                                                                     960

 P  M  I  I  I  R  E  G  L  D  H  Q  D  R  W  E  W  F  I  D  Y  M  D  R  K  H  T  C  Q  L  C  Y  L  K  H  S  K  Q  I  P
ACCTATGATCATCATTCGTGAGGGGTTGGATCATCAGGATCGGTGGGAATGGTTTATTGATTACATGGATAGGAAACATACATGTCAACTTTGCTACTTGAAACATTCAAAACAGATCCC
                                                                                                                    1080

 T  C  S  V  I  D  V  R  A  S  E  L  T  G  C  S  P  F  K  M  V  K  I  E  E  H  V  G  N  D  S  V  F  K  T  K  L  V  R  D
AACCTGTAGTGTGATTGATGTACGTGCATCTGAATTAACTGGGTGCTCGCCGTTCAAGATGGTGAAGATCGAAGAGCATGTAGGAAATGATTCAGTGTTTAAAACGAAATTAGTTCGCGA
                                                                                                                    1200

 E  Q  I  G  R  I  G  D  H  Y  Y  T  T  N  C  Y  T  G  A  E  A  L  I  T  T  A  I  H  I  H  H  W  I  R  G  S  G  I  W  N
TGAGCAAATCGGCAGGATTGGAGATCATTATTATACAACAAATTGTTACACTGGAGCGGAGGCGTTGATTACAACCGCGATTCACATTCATCACTGGATTAGGGGGTCTGGCATCTGGAA
                                                                                                                    1320

 D  E  G  W  Q  E  G  I  F  M  L  G  R  V  L  L  R  W  E  L  T  K  A  Q  R  S  A  L  L  R  L  F  C  F  V  C  Y  G  Y  A
CGATGAAGGATGGCAGGAGGGTATTTTCATGCTTGGACGCGTGCTGCTGAGATGGGAATTGACAAAGGCGCAACGCAGCGCTTTGCTCAGGCTATTCTGTTTTGTATGTTACGGATATGC
                                                                                                                    1440

 P  R  A  D  G  T  I  P  D  W  N  N  L  G  N  F  L  D  I  I  L  K  G  P  E  L  S  E  D  E  D  E  R  A  Y  A  T  M  F  E
ACCACGCGCAGACGGAACGATACCGGACTGGAATAATCTTGGAAACTTTTTGGATATCATTTTGAAGGGGCCAGAACTTAGTGAAGATGAGGATGAAAGAGCTTATGCTACAATGTTTGA
                                                                                                                    1560

 M  V  R  C  I  I  T  L  C  Y  A  E  K  V  H  F  A  G  F  A  A  P  A  C  E  G  G  E  V  I  N  L  A  A  A  M  S  Q  M  W
AATGGTTCGATGCATTATCACTCTATGCTATGCAGAAAAGGTTCACTTCGCTGGGTTCGCTGCGCCTGCGTGTGAAGGCGGGGAAGTAATTAATCTTGCTGCGGCCATGTCTCAGATGTG
                                                                                                                    1680

 M  E  Y  *
GATGGAGTATTAGTTACTGATTTTTAGTTTTTTATTCTTCTTTTCATTTCTATTTCTCTTAGCACTCTACTAGAACTTTTCAACTTAC
                                                                                       1769
```

S7

```
                 M  D  T  I  A  A  R  A  L  T  V  M  R  A  C  A  T  L  Q  E  A  R  I  V  L  E  A  N  V  M  E  I  L  G
GTTAAAAATCTATAGAGATGGACACTATCGCCGCAAGAGCACTCACTGTGATGCGAGCATGTGCTACGCTTCAAGAGGCAAGAATTGTGTTGGAAGCCAATGTGATGGAAATTTTGGGGA
                                                                                                                     120

I  A  I  N  R  Y  N  G  L  T  L  R  G  V  T  M  R  P  T  S  L  A  Q  R  N  E  M  F  F  M  C  L  D  M  M  L  S  A  A  G
TAGCTATCAATAGGTACAATGGACTCACTTTACGAGGAGTGACGATGCGCCCGACCTCGTTAGCACAAAGAAATGAGATGTTTTTTATGTGTTTGGATATGATGCTGTCTGCTGCTGGGA
                                                                                                                     240

I  N  V  G  P  I  S  P  D  Y  T  Q  H  M  A  T  I  G  V  L  A  T  P  E  I  P  F  T  T  E  A  A  N  E  I  A  R  V  T  G
TAAATGTTGGACCGATATCGCCAGACTATACTCAACATATGGCTACGATTGGTGTACTAGCAACACCGGAAATACCTTTTACAACGGAAGCGGCGAATGAAATAGCTCGAGTGACTGGGG
                                                                                                                     360

E  T  S  T  W  G  P  A  R  Q  P  Y  G  F  F  L  E  T  E  E  T  F  Q  P  G  R  W  F  M  R  A  A  Q  A  V  T  A  V  V  C
AGACTTCGACATGGGGGCCAGCGCGTCAGCCTTATGGTTTCTTCCTTGAAACTGAGGAAACCTTCCAACCAGGGAGGTGGTTCATGCGCGCCGCTCAAGCAGTAACTGCAGTAGTGTGCG
                                                                                                                     480

G  P  D  M  I  Q  V  S  L  N  A  G  A  R  G  D  V  Q  Q  I  F  Q  G  R  N  D  P  M  M  I  Y  L  V  W  R  R  I  E  N  F
GTCCGGATATGATTCAAGTGTCACTTAATGCTGGAGCGAGAGGAGATGTACAACAGATATTTCAGGGTCGTAATGATCCCATGATGATATATTTAGTGTGGAGGAGAATCGAAAACTTTG
                                                                                                                     600

A  M  A  Q  G  N  S  Q  Q  T  Q  A  G  V  T  V  S  V  G  G  V  D  M  R  A  G  R  I  I  A  W  D  G  Q  A  A  L  H  V  H
CGATGGCGCAAGGTAATTCACAGCAAACTCAAGCGGGTGTGACTGTCAGTGTTGGTGGAGTTGACATGAGGGCGGGACGCATTATAGCGTGGGATGGACAGGCCGCGCTGCATGTGCATA
                                                                                                                     720

N  P  T  Q  Q  N  A  M  V  Q  I  Q  V  V  F  Y  I  S  M  D  K  T  L  N  Q  Y  P  A  L  T  A  E  I  F  N  V  Y  S  F  R
ATCCGACACAACAGAATGCGATGGTGCAAATACAGGTTGTGTTCTATATATCTATGGATAAAACTTTAAACCAGTACCCCGCTTTGACTGCTGAGATTTTCAATGTTTACAGCTTCAGGG
                                                                                                                     840

D  H  T  W  H  G  L  R  T  A  I  L  N  R  T  T  L  P  N  M  L  P  P  I  F  P  P  N  D  R  D  S  I  L  T  L  L  L  L  S
ACCACACATGGCATGGGCTAAGAACGGCGATATTAAACAGAACCACACTGCCAAACATGCTGCCACCAATCTTCCCACCAAATGATCGAGATAGCATCTTAACTCTTCTACTTTTATCTA
                                                                                                                     960

T  L  A  D  V  Y  T  V  L  R  P  E  F  A  I  H  G  V  N  P  M  P  G  P  L  T  R  A  I  A  R  A  A  Y  V  *
CACTTGCTGATGTTTACACTGTTTTAAGGCCAGAGTTTGCGATTCACGGCGTAAATCCGATGCCAGGGCCGCTCACACGTGCTATTGCGCGCGCCGCCTATGTGTAGTCCACTTTGCACG
                                                                                                                    1080

GGTGTGGGTTACATATGCGGTGTGTCGGTTGTGGGAAATATGTAACCCATTTAAACGTCTCTTAGATTACACTTAC
                                                                        1156
```

S8

```
                    M  E  Q  K  Q  R  R  F  T  K  N  I  F  V  L  D  V  T  A  K  T  L  C  G  A  I  A  K  L  S  S  Q  P  Y
GTTAAAAAATCCTTGAGTCATGGAGCAAAAGCAACGTAGATTTACCAAAAACATTTTGTCTTGGACGTAACCGCTAAAACATTATGCGGGGCTATCGCGAAGTTGAGTTCGCAGCCGTA
                                                                                                                     120

  C  Q  I  K  I  G  R  V  V  A  F  K  P  V  K  N  P  E  P  K  G  Y  V  L  N  V  P  G  P  G  A  Y  R  I  Q  D  G  Q  D  I
TTGTCAAATTAAAATTGGAAGAGTAGTAGCTTTTAAACCTGTCAAGAATCCGGAACCTAAAGGATACGTGCTGAATGTTCCAGGACCAGGTGCATACAGAATTCAGGATGGGCAGGATAT
                                                                                                                     240

  I  S  L  M  L  T  P  H  G  V  E  A  T  T  E  R  W  E  E  W  K  F  E  G  V  S  V  T  P  M  A  T  R  V  Q  Y  N  G  V  M
CATTAGCTTGATGCTGACGCCACATGGAGTCGAAGCGACAACGGAAAGATGGGAAGAGTGGAAATTTGAGGGTGTTAGCGTGACGCCAATGGCTACGCGGGTACAGTATAATGGTGTGAT
                                                                                                                     360

  V  D  A  E  I  K  Y  C  K  G  M  G  I  V  Q  P  Y  M  R  N  D  F  D  R  N  E  M  P  D  L  P  G  V  M  R  S  N  Y  D  I
GGTTGACGCTGAAATTAAATACTGTAAGGGGATGGGAATTGTGCAACCATATATGCGGAATGATTTTGATCGAAATGAAATGCCCGATTTGCCAGGCGTGATGAGATCAAATTATGATAT
                                                                                                                     480

  R  E  L  R  Q  K  I  K  N  E  R  E  S  A  P  R  L  Q  V  H  S  V  A  P  R  E  E  S  R  W  M  D  D  D  E  A  K  V  D  D
CCGCGAATTACGGCAAAAGATCAAAAATGAACGAGAATCAGCGCCACGGCTTCAAGTTCATAGCGTGGCGCCAAGGGAGGAGTCGCGCTGGATGGATGATGATGAGGCGAAGGTGGATGA
                                                                                                                     600

  E  A  K  E  I  V  P  G  T  S  G  L  E  K  L  R  E  A  R  S  N  V  F  K  E  V  E  A  V  I  N  W  N  L  D  E  R  D  E  G
TGAGGCCAAAGAGATAGTTCCGGGAACTAGCGGGTTGGAAAAACTACGTGAGGCGAGAAGCAATGTTTTTAAAGAGGTGGAAGCTGTAATTAATTGGAATCTAGATGAGAGAGATGAAGG
                                                                                                                     720

  D  R  D  E  R  G  D  E  E  Q  V  K  T  L  S  D  D  D  D  Q  G  E  D  A  S  D  D  E  H  P  K  T  H  I  T  K  E  Y  I  E
GGATAGAGATGAACGGGGGGATGAGGAGCAAGTGAAGACCTTGAGTGACGATGATGACCAAGGCGAGGATGCGAGTGATGATGAACACCCAAAAACCCATATAACGAAAGAGTATATCGA
                                                                                                                     840

  K  V  A  K  Q  I  K  L  K  D  E  R  F  M  S  L  S  S  A  M  P  Q  A  S  G  G  F  D  R  M  I  V  T  K  K  L  K  W  Q  N
AAAAGTTGCAAAACAGATTAAATTGAAAGATGAACGGTTCATGAGTCTATCAAGCGCTATGCCTCAAGCGAGTGGTGGATTTGATCGTATGATTGTTACGAAGAAGCTTAAGTGGCAGAA
                                                                                                                     960

  V  P  L  Y  C  F  D  E  S  L  K  R  Y  E  L  Q  C  V  G  A  C  E  R  V  A  F  V  S  K  D  M  S  L  I  I  C  R  S  A  F
TGTGCCACTGTATTGCTTTGATGAGTCATTGAAGAGGTATGAGTTGCAGTGTGTGGGTGCGTGTGAGCGTGTTGCTTTTGTCTCTAAGGACATGAGCTTGATCATATGCCGGTCGGCGTT
                                                                                                                    1080

  R  R  L  *
TAGGCGCTTGTGACCGCATGATTGGGGGGGGATTTTACACTTAC
                                        1124
```

S9

```
                M  S  A  A  I  L  L  A  P  G  D  V  I  K  R  S  S  E  E  L  K  Q  R  Q  I  Q  I  N  L  V  D  W  M  E  S
GTTAAAAAATCGCATATGTCAGCTGCGATACTTCTTGCACCCGGTGACGTGATCAAGCGTTCATCCGAGGAGTTAAAACAGAGGCAGATTCAGATTAATTTGGTTGATTGGATGGAGAGT
                                                                                                                     120

 E  G  G  K  E  D  K  T  E  P  K  E  E  S  K  A  E  G  S  K  D  G  E  G  R  N  R  R  A  A  R  R  K  R  A  G  K  E  T  K
GAGGGCGGAAAAGAGGATAAAACAGAACCTAAAGAGGAGAGCAAAGCAGAAGGATCAAAGGATGGTGAGGGACGCAATCGGAGAGCGGCCAGAAGGAAGAGGGCGGGCAAAGAGACTAAA
                                                                                                                     240

 D  A  E  C  D  R  R  I  H  T  A  V  G  S  G  S  G  T  K  G  S  G  E  R  A  N  E  N  A  N  R  G  D  G  K  V  G  G  G  G
GATGCAGAATGTGATAGACGCATACATACTGCAGTGGGATCTGGATCAGGCACAAAAGGATCTGGAGAACGCGCGAACGAGAATGCTAACAGAGGAGATGGAAAGGTTGGAGGAGGAGGT
                                                                                                                     360

 G  D  A  D  A  G  V  G  A  T  G  T  N  G  G  R  W  V  V  L  T  E  E  I  A  R  A  I  E  S  K  Y  G  T  K  I  D  V  Y  R
GGAGATGCTGATGCGGGAGTTGGAGCTACTGGAACGAATGGAGGAAGATGGGTAGTTCTAACGGAAGAGATTGCTCGCGCGATTGAATCTAAATACGGTACGAAGATTGATGTTTACAGG
                                                                                                                     480

 D  D  V  P  A  Q  I  I  E  V  E  R  S  L  Q  K  E  L  G  I  S  R  E  G  V  A  E  Q  T  E  R  L  R  D  L  R  R  K  E  K
GATGACGTTCCAGCTCAGATCATCGAGGTGGAACGCAGTCTCCAGAAAGAGTTGGGAATTTCCCGTGAGGGAGTGGCCGAGCAGACCGAGAGATTGAGAGATCTGCGGCGGAAAGAGAAG
                                                                                                                     600

 N  G  T  H  A  K  A  V  E  R  G  G  R  K  Q  R  K  K  A  H  G  D  A  Q  R  E  G  V  E  E  E  K  T  S  E  E  P  A  R  I
AATGGGACGCATGCTAAAGCCGTTGAGAGAGGAGGACGCAAGCAGAGGAAGAAAGCCCATGGTGACGCGCAGAGAGAAGGTGTCGAGGAAGAAAAGACGTCTGAGGAGCCGGCACGCATT
                                                                                                                     720

 G  I  T  I  E  G  V  M  S  Q  K  K  L  L  S  M  I  G  G  V  E  R  K  M  A  P  I  G  A  R  E  S  A  V  M  L  V  S  N  S
GGAATTACTATAGAAGGGGTCATGAGTCAAAAGAAACTGCTAAGCATGATTGGTGGTGTGGAGAGGAAGATGGCTCCGATAGGAGCGCGGGAGAGCGCGGTTATGTTAGTTTCAAACAGT
                                                                                                                     840

 I  K  D  V  V  R  A  T  A  Y  F  T  A  P  T  G  D  P  H  W  K  E  V  A  A  K  L  R  K  R  R  N  I  S  Y  T  S  T  G  G
ATAAAAGATGTGGTGCGAGCAACGGCATATTTCACAGCGCCAACTGGAGATCCGCATTGGAAAGAAGTGGCTGCGAAGCTTCGAAAAAGAAGAAATATTAGTTATACGAGTACAGGGGGT
                                                                                                                     960

 D  V  K  T  E  F  L  H  L  I  D  H  L  *
GATGTGAAAACGGAGTTTCTACATTTGATCGATCACCTCTAAAGGGTCCAGGTACCTTCTTGACGTAGGGCGATTTCACACTTAC
                                                                                   1046
```

S10

```
                    M  L  S  G  L  I  Q  R  F  E  E  E  K  M  K  H  N  Q  D  R  V  E  E  L  S  L  V  R  V  D  D  T  I  S
GTTAAAAAGTGTCGCTGCCATGCTATCCGGGCTGATCCAAAGGTTCGAAGAAGAAAAGATGAAACACAATCAAGATAGAGTTGAAGAACTGAGTTTAGTGCGCGTGGATGATACCATTTC
                                                                                                                     120

  Q  P  P  R  Y  A  P  S  A  P  M  P  S  S  M  P  T  V  A  L  E  I  L  D  K  A  M  S  N  T  T  G  A  T  Q  T  Q  K  A  E
TCAACCGCCAAGGTATGCTCCGAGTGCACCGATGCCATCATCCATGCCAACGGTTGCCCTTGAAATACTGGACAAAGCGATGTCAAACACAACTGGTGCAACGCAAACACAAAAGGCGGA
                                                                                                                     240

  K  A  A  F  A  S  Y  A  E  A  F  R  D  D  V  R  L  R  Q  I  K  R  H  V  N  E  Q  I  L  P  K  L  K  S  D  L  S  E  L  K
GAAGGCTGCATTCGCATCGTACGCAGAAGCGTTTCGTGATGACGTAAGACTAAGACAGATTAAACGCCATGTAAATGAACAGATTTTACCTAAATTGAAAAGTGATCTAAGTGAATTGAA
                                                                                                                     360

  K  K  R  A  I  I  H  T  T  L  L  V  A  A  V  V  A  L  L  T  S  V  C  T  L  S  S  D  M  S  V  A  F  K  I  N  G  T  K  T
GAAGAAGCGAGCAATCATACACACTACTCTACTAGTGGCTGCTGTGGTTGCGCTGCTGACATCAGTTTGCACCCTTTCAAGTGATATGAGTGTGGCCTTCAAAATAAATGGGACCAAAAC
                                                                                                                     480

  E  V  P  S  W  F  K  S  L  N  P  M  L  G  V  V  N  L  G  A  T  F  L  M  M  V  C  A  K  S  E  R  A  L  N  Q  Q  I  D  M
AGAAGTGCCTTCATGGTTTAAAAGCCTTAATCCAATGCTTGGCGTTGTCAACTTGGAGCAACTTTTTTGATGATGGTTTGCGCAAAGAGTGAAAGAGCCTTGAATCAACAGATAGATAT
                                                                                                                     600

  I  K  K  E  V  M  K  K  Q  S  Y  N  D  A  V  R  M  S  F  T  E  F  S  S  I  P  L  D  G  F  E  M  P  L  T  *
GATAAAGAAAGAAGTGATGAAAAAACAATCATATAATGACGCAGTGAGGATGAGTTTTACAGAGTTCTCGTCGATCCCGCTGGATGGTTTTGAAATGCCATTAACCTGAGGACAGTAGGT
                                                                                                                     720

AGAGTGGCGCCCCAAGGTTTACGTCGTGCAGGGTGGTTGACCTCGCGGCGTAAATTCCCACTGCTGTATAACGGGGGAGGGTGCGCGATACTACACACTTAC
                                                                                                    822
```

Subject Index

Abadina 10
acridine orange 104
adsorption to cells 93
African horse-sickness 5
agar gel immunodiffusion tests 4
amino acid composition 56
amplification 96, 114
antibodies 27
  group-specific 29
  monoclonal antibodies 27, 28
  serotype-specific 27, 28
antigenic determinants 69
antigenic variation 59
aromatic residues 56, 57

Baculovirus, polyhedrin 71
bluetongue 2
  in Australia 3
  in the Caribbean and Central America 15
  in Cyprus 2
  in India 3
  in Israel 2, 3
  in Portugal 2
  in Spain 2
  in Turkey 2
  in the United States 2
  in West Pakistan 3
blutongue virus, populations 13
  prototype strains 6
BTV, barriers to infection 155
  Culicoides 155, 156
  digestive enzymes 156, 157
  other anthropods 153, 154
BTV in blood 126
  antibody prevalence 129
  buffy coat 126
  clinical disease 128
    in cattle 130
  clinically infected animals 130
  endemic 130
    areas 128
  epizootics 128, 130
  erythrocyte 126
    fractions 126
  erythrocyte-granulocyte fractions 126
  immune responses to VP2 127
  mononuclear fractions 126
  neutralizing antibodies 127, 128
  neutralizing reactions 128
  prevalence of serum antibodies 130
  proteins NS1 and NS2 127
  serological evidence 130
  serological survey 130
BTV, occurrence
  Africa and the Middle East 145
  America 147
  Asia 145, 146
  Australia 146, 147
  Europe 147, 148
BTV, particle types
  core particles 156, 157
  infections subviral particles 156, 157
  intact particles 156, 157
BTV serotypes 164

Capsid proteins (see structural proteins)
capsomeres 23, 29
cattle 170
cell attachment 28
cell-mediated immune responses 169, 172, 174
  ^{51}Cr-release assays 174
  major histocompatibility complex-restricted, cytotoxic T-lymphocytes 174
cell viability 112
cellular immune response 166
  cytotoxicity 166
  lymphocyte blastogenesis 166
clathrin 94
clinical disease 167
clinical signs 121
  clinical disease cattle 128–130
  clinical disease other species 134

clinical disease sheep 121
clinical pathology 123
hermorrhages 121
coding assignments (see also double-stranded RNA) 24
colchicine 102
common ancestor 70
complement-fixation (CF) tests 4
complete sequence of the BTV-10 genome 54, 70
base composition 54
3'noncoding regions 54
5'noncoding regions 54
conserved amino acid sequences 65, 69
conserved regions 59
core particles 22
association with nonstructural proteins 30
composition 29
CsCl density 23
diameter 23
isolation and purification 23, 35
morphological features 23
mRNA capping activity 36
stability 24
transcriptase activity 35, 36
cross-challenge 164
cross-immune precipitation experiments 8
cross-protection tests 2
cross-reactivity 165
cryoelectron microscopy 75
Culicoides vectors, C. acetoni 146, 147
C. fulvus 145–147
C. imicola 145–148
C. nubeculosus 148
C. variipennis 147–153, 155–157
C. wadai 145–147
cytochalasin B 102
cytohybridization 13
cytoskeleton 101–108
cytoskeleton-associated virus 115
particles 107
cytosol virus particles 107, 115

Diagnosis 76
indirect ELISA 77, 81
reagent 81
Diagon analyses 59, 66
diplornaviruses 4
disulfide bridges 59, 68
DNA clones 48
RNA-cDNA hybrid duplexes 48
double stranded RNA (dsRNA) 4
cloning and sequencing 32, 37
coding assignments of individual segments 24
electrophoretic migration of 32
genomic probes 34
heterogeneity 34
linkage inside virus particle 33
noncoding regions 32
nucleic acid homology (similarity) 27, 29, 34
physicochemical characteristics 32
profiles 32, 33
dsRNA (double-stranded RNA) 4
dsRNA Segments 44, 54
coding arrangements 54
3'terminal sequences 48

EHD (epizootic hemorrhagic disease) 5
EHD virus infection 135
buffalo 135
cattle 135
deer 135
wild ruminant 135
ELISA (enzyme-linked immunsorbent assay) 7, 8
blocking 12
competitive 12
indirect 12
electronmicroscopy 22
core particles 22
subcore particles 22
tubules 30
virions 22
electrophoretic migration (see also viral proteins, double-stranded RNA and mRNA) 24, 32
endocytosis, bluetongue virus 93, 94, 114
reovirus 91
endosome 94, 96, 114
enveloped viruses 111
enveloped virus particles 108, 112
enzyme-linked immunsorbent assay (ELISA) 7, 8
epidemiology, endemic areas of Australia 127
epitopes on VP2 165
immunoprecipitation 165
reassortant viruses 165
epizootic hemorrhagic disease (EHD) 5
virus 165
prototype strains 9
serogroup 8
XBM/67 8
epizootiology 128

antibody prevalence 129, 130
endemic 120, 130
epidemiological studies 127
epizootie 120, 128, 130
serological survey 130
Eubenangee 5, 9
virus serogroup 9

Fetal immune responses 170, 173
cerebral malformation 173
CMI 173
immunological tolerance 173
immunologically unresponsive 173
lymphocyte blastogenesis 173
persistently infected 173
fluorescent-antibody tests 4

Gene pools 13
gene reassortment 9
genetic relationships 48
between different serotypes 48
Northern blot analysis 49
genomic probes 34
conserved 34
for predominant mRNA species 34
serogroup-specific 34
serotype-specific 34
gold labelled antibodies 102–111
gold-silver staining 109
grid cell culture technique 106, 108
group-specific antigen 68, 83
group-specific antigenic determinants 62

Hemagglutination 7
hemagglutinin protein 57
hemolysis in gel (HIG) test 7
humoral immune responses 127, 164, 170
humoral immunity, neutralizing antibodies 127, 131–133
non-structural protein 127
serum antibodies 133

Ibaraki virus 8
immune-mediated pathogenesis 172
BTV-specific IgE 172
dermatitis 172
immediate type hypersensitivity (type I) 172
passive cutaneous anaphylaxis (PCA) 172
reagenic antibody 172
stomatitis 172
immune responses (see also antibodies) 163
analysis by immuneprecipitation 27
cell-mediated 37
group-specific 29
humoral 37
of mice 173
intracerebral challenge 174
monoclonal antibodies 174
nonneutralizing antibody 174
passive transfer 174
virus-neutralizing antibodies 174
neutralising antibodies 27, 28
protective 27, 37
serotype-specific 29
immunization, with purified VP2 165
with recombinant VP2 165
reassortant viruses 165
immunoelectron microscopy 81, 102–111
immunofluorescence 98–101, 104, 105, 108
immunogold 75
immunological techniques 164
immunopotentiators 172
in vitro translation 92
of RNA segments 9
indirect immunoperoxidase technique 12
infection EHDV 135
cattle 135
wild ruminants 135
infections of culicoides, oral infections 150–152
parenteral inoculation 148, 149
persistence of infection 152
variations in susceptibility 152, 153
intermediate filaments 101, 102
ionophores, nigericin 96, 98

Localization of virus proteins, infected cells
NS1 103, 105, 106, 114, 115
NS2 105
VP2 101, 104–115
VP3 103, 104, 115
VP7 103, 104, 115
localization of VP7 in virus particles 92
lysomotropic 114
weak bases, NH_4Cl 96, 98, 100

Messenger RNA (single-stranded RNA) 35
capping 36
electrophoretic fractionation 35, 36

relative amounts synthesised 30, 35
S value 35
synthesis 35
monoclonal antibodies 90, 99–102
immuno-cytochemistry 90, 98–115
morphogenesis 104–113
mutants, neutralization-specific escape mutation 28

Neutralizing antigen VP2 57
nonstructural proteins (see also viral proteins)
affinity for single-stranded RNA 31
association with tubules 30, 31
coding segment 24, 30, 31
function of 31
genes coding for 30, 31
phosphorylated 31
purification of 31
relative amount synthesised 30, 31
nucleic acid hybridisation 27
dsRNA/dsRNA 27, 28
mRNA/dsRNA 28, 36
northern blots 27
NS2 81

Oligonucleotide fingerprint 46, 70
evolution of the BTV 11, 47
oligonucleotide maps 13
Orbivirus 4
serological groups 5
African horse sickness 5
bluetongue 5
Changuinola 5
Corriparta 5
epizootic hemorrhagic disease 5
equine encephalosis 5
Eubenangee 5
Kemerovo 5
Palyam 5
Umatilla 5
Wallal 5
Warrego 5

Palyam virus serogroup 10
parturition-associated immunosuppression 170
passive hemagglutination (PHA) test 7
Pata virus 10
pathogenesis
antigen concentration 125
Culicoides sp. 125
endothelial cells 124, 126
erythrocytes 126
histological lesions 125
immunofluorescence 124, 125
lymphoid tissues 125
lymphoreticular system 125
natural transmission 125
periendothelial cells 124
skeletal muscle 124
in sheep 124
vascular endothelium 125
viral antigen 124, 125
pathology 122
cardiovascular 122
clinical 123
digestive tract 122
external lesions 122
hemorrhages 122
and congestion 122
ecchymotic 122
histopathological 123–125
microscopic lesions 123
microvascular 123
respiratory system 122
in sheep 122
skeletal muscle 124
vascular system 121–125
vascular thrombosis 123
phophoprotein 81
protective immunity 167, 169, 171, 172
cell-mediated immune (CMI) 169
CMI 172
cross-serotype reactivity 169
cytotoxicity 169
cytotoxic T lymphocytes 169
heterotypic virus-neutralizing antibody 171
immunological responsiveness 169
lymphocyte blastogenesis 172
neutralizing antibody 171
Suffolk-cross sheep 169
transfer of such immune lymphocytes 169
Warhill sheep 169
protein A-gold 105

Quarantine 3

Reassortant BTVs 13
reassortant viruses 28
reassortment of genome segments 47
in Culicoides vectors 47
in tissue culture 47

in vertebrate hosts 47
recombinant baculoviruses 70
release 108
by extrusion 112, 113
reovirus 94
inclusion bodies 101
morphogenesis 91
replication 90
reproduction
abortions 131, 132
artificial breeding centers 133
congenital defects 131, 132
embryo 132, 133
embryo transfer 131–133
fetal abnormalities 131, 133
transplacental 132
reproduction in cattle 130
abortigenic 133
abortions 132
congenital 132
defects 132
embryo transfer 133
humoral immune response 132
neutralizing antibodies 132, 133
semen 133
serum antibodies 133
transplacental 132
reproduction in sheep 130
congenital defects 131
embryos 131
fetal abnormalities 131
serum-neutralizing antibodies 131
RNA-binding protein 78
RNA polymerase (transcriptase) 21
activation 21, 35
common tobacco chloroplast 56
control 36
Drosophila polymerase 56
Escherichia coli 56
inhibitory effect 35
RNA transcription 29
Saccharonyces cerevisiae 56
temperature optimum 35
RNA-RNA hybridization 8
RNA-RNA reassociation 13
RNA segments, in vitro translation 9

Serogroup specific 164
antibody 165, 166, 170
agar gel immunodiffusion 165
AGID 165, 170
CF 165
colostrum-deprived 171
complement-fixation (CF) 165
ELISA 165, 171
enzyme-linked immunosorbent assay 165
hemolysis in gel 165
immunoblotting 166
immunoprecipitation 166, 171
loss of AGID antibody 171
serotype-determining antigen 57
serotype-specific 164
antibody 164, 166, 171
cytopathic effect 164
hemagglutination inhibition 164
HI antibody 165
monoclonal antibody 164
neutralizing 165
passive protection 164
plaque-reduction 164
protein specificity 165
virus-neutralizing antibody 164, 171
virus-neutralization assays 164
VP2 165
VP5 165
serotyping 2
serum-neutralization tests 2
sheep 166
protective immunity 167
AGID 167
cell-mediated lysis 167
cellular 167
CF 167
colostral 167
complement-mediated lysis 167
heterotypic neutralizing antibody 168
humoral 167
monoclonal antibody 168
passive transfer 167
serogroup-specific neutralizing antibody 168
virus-neutralizing antibody 167
serogroup-specific antibody 166
AGID 166
CF 166
ELISA 166
immunoblot 166
immunoprecipitation 166
serotype-specific antibody 166
virus-neutralizing antibody 166
soremuzzle 2
Spodoptera frugiperda, cells 68, 72
stoichiometry 76
structural proteins (see also viral proteins) 36
association with polymerase activity 36
coding segment 24
core particle 25, 26, 29, 36

enzymatic function 29, 36
estimated number/virion 24, 26
group-specific antigen 29
heamagglutination activity 28, 29
location 24
major and minor proteins 25, 26, 29
mRNA capping activity 36
neutralisation-specific immune response 27, 28
outer capsid layer 25–29
percentage of total in virion 24
subcore particles 23
composition 23
diameter 23
major structural protein 29
morphology and appearance 22, 23
purification 23
subunite vaccine 73
cross-neutralization 53, 73
superinfection 98

Tilligerry virus 10
transfer vector 71
tubules 78, 100, 102, 103, 106, 115
BTV-infected 78
recombinant virus 78

Umatilla virus serogroup 9
Llano Seco 9
Minnal 9
Netivot 9
Umatilla 9

Vaccine 2
vectors 120
culicoides brevitarsis 120
culicoides variipennis 129, 131
endemic 120
epizootic 120
serum antibodies 120
natural transmission 125
species Culicoides 120
VIB 100, 102, 105, 106, 114
vimentin 102
viral nucleic acids (see double-stranded RNA and mRNA)
viral proteins (see also structural proteins and nonstructural proteins)
amino acid sequences 27
coding assignments 24, 25
core polypeptides 29
electrophoretic migration (PAGE) 24–26, 29
estimated number/virion 24
expression 30
molecular mass 24, 25
outer capsid 26–29
peptide mapping 26, 29, 31
relative frequency synthesised 26
variability 26, 27
virion replicase/transcriptase 77
virions 22
association with nonstructural proteins 22
buoyant density 23
chemical composition 23
diameter 23
infectivity 24
morphology 22
purification 22, 23
number of estimated proteins/virion 24
removal of outer capsid layer 23, 35
sedimentation constant 23
stability 23, 24
uncoating 23
virulence 28
virus clearance 168, 171, 172
AGID 171
CMI 170, 172
coexistence of virus and neutralizing antibody 171
CTL 170
immunoblot 171
immunoprecipitation 171
lymphocyte blastogenesis 172
neutralizing antibody 171
virus-antibody complexes 168
virus inclusion bodies 97, 104
virus neutralization 73
virus purification 22
virus tubules 30
characteristics 30
diameter 30
localisation 30
polymerisation of 30
sedimentation coefficient 30
VP6 78
VP7 53

Western immunoblotting 12

Current Topics in Microbiology and Immunology

Volumes published since 1983 (and still available)

Vol. 107: **Vogt, Peter K.; Koprowski; Hilary (Ed.):** Retroviruses 2. 1983. 26 figs. VII, 180 pp. ISBN 3-540-12384-9

Vol. 109: **Doerfler, Walter (Ed.):** The Molecular Biology of Adenoviruses 1. 30 Years of Adenovirus Research 1953–1983. 1983. 69 figs. XII, 232 pp. ISBN 3-540-13034-9

Vol. 110: **Doerfler, Walter (Ed.):** The Molecular Biology of Adenoviruses 2. 30 Years of Adenovirus Research 1953–1983. 1984. 49 figs. VIII, 265 pp. ISBN 3-540-13127-2

Vol. 112: **Vogt, Peter K.; Koprowski, Hilary (Ed.):** Retroviruses 3. 1984. 19 figs. VII, 115 pp. ISBN 3-540-13307-0

Vol. 113: **Potter, Michael; Melchers, Fritz; Weigert, Martin (Ed.):** Oncogenes in B-Cell Neoplasia. Workshop at the National Cancer Institute, National Institutes of Health, Bethesda, MD March 5–7, 1984. 1984. 65 figs. XIII, 268 pp. ISBN 3-540-13597-9

Vol. 115: **Vogt, Peter K. (Ed.):** Human T-Cell Leukemia Virus. 1985. 74 figs. IX, 266 pp. ISBN 3-540-13963-X

Vol. 116: **Willis, Dawn B. (Ed.):** Iridoviridae. 1985. 65 figs. X, 173 pp. ISBN 3-540-15172-9

Vol. 122: **Potter, Michael (Ed.):** The BALB/c Mouse. Genetics and Immunology. 1985. 85 figs. XVI, 254 pp. ISBN 3-540-15834-0

Vol. 124: **Briles, David E. (Ed.):** Genetic Control of the Susceptibility to Bacterial Infection. 1986. 19 figs. XII, 175 pp. ISBN 3-540-16238-0

Vol. 125: **Wu, Henry C.; Tai, Phang C. (Ed.):** Protein Secretion and Export in Bacteria. 1986. 34 figs. X, 211 pp. ISBN 3-540-16593-2

Vol. 126: **Fleischer, Bernhard; Reimann, Jörg; Wagner, Hermann (Ed.):** Specificity and Function of Clonally Developing T-Cells. 1986. 60 figs. XV, 316 pp. ISBN 3-540-16501-0

Vol. 127: **Potter, Michael; Nadeau, Joseph H.; Cancro, Michael P. (Ed.):** The Wild Mouse in Immunology. 1986. 119 figs. XVI, 395 pp. ISBN 3-540-16657-2

Vol. 128: 1986. 12 figs. VII, 122 pp. ISBN 3-540-16621-1

Vol. 129: 1986. 43 figs., VII, 215 pp. ISBN 3-540-16834-6

Vol. 130: **Koprowski, Hilary; Melchers, Fritz (Ed.):** Peptides as Immunogens. 1986. 21 figs. X, 86 pp. ISBN 3-540-16892-3

Vol. 131: **Doerfler, Walter; Böhm, Petra (Ed.):** The Molecular Biology of Baculoviruses. 1986. 44 figs. VIII, 169 pp. ISBN 3-540-17073-1

Vol. 132: **Melchers, Fritz; Potter, Michael (Ed.):** Mechanisms in B-Cell Neoplasia. Workshop at the National Cancer Institute, National Institutes of Health, Bethesda, MD, USA, March 24–26, 1986. 1986. 156 figs. XII, 374 pp. ISBN 3-540-17048-0

Vol. 133: **Oldstone, Michael B. (Ed.):** Arenaviruses. Genes, Proteins, and Expression. 1987. 39 figs. VII, 116 pp. ISBN 3-540-17246-7

Vol. 134: **Oldstone, Michael B. (Ed.):** Arenaviruses. Biology and Immunotherapy. 1987. 33 figs. VII, 242 pp. ISBN 3-540-17322-6

Vol. 135: **Paige, Christopher J.; Gisler, Roland H. (Ed.):** Differentiation of B Lymphocytes. 1987. 25 figs. IX, 150 pp. ISBN 3-540-17470-2

Vol. 136: **Hobom, Gerd; Rott, Rudolf (Ed.):** The Molecular Biology of Bacterial Virus Systems. 1988. 20 figs. VII, 90 pp. ISBN 3-540-18513-5

Vol. 137: **Mock, Beverly; Potter, Michael (Ed.):** Genetics of Immunological Diseases. 1988. 88 figs. XI, 335 pp. ISBN 3-540-19253-0

Vol. 138: **Goebel, Werner (Ed.):** Intracellular Bacteria. 1988. 18 figs. IX, 179 pp. ISBN 3-540-50001-4

Vol. 139: **Clarke, Adrienne E.; Wilson, Ian A. (Ed.):** Carbohydrate-Protein Interaction. 1988. 35 figs. IX, 152 pp.
ISBN 3-540-19378-2

Vol. 140: **Podack, Eckhard R. (Ed.):** Cytotoxic Effector Mechanisms. 1989. 24 figs. VIII, 126 pp. ISBN 3-540-50057-X

Vol. 141: **Potter, Michael; Melchers, Fritz (Ed.):** Mechanisms in B-Cell Neoplasia 1988. Workshop at the National Cancer Institute, National Institutes of Health, Bethesda, MD, USA, March 23–25, 1988. 1988. 122 figs. XIV, 340 pp.
ISBN 3-540-50212-2

Vol. 142: **Schüpbach, Jörg:** Human Retrovirology. Facts and Concepts. 1989. 24 figs. 115 pp. ISBN 3-540-50455-9

Vol. 143: **Haase, Ashley T.; Oldstone, Michael B. A. (Ed.):** In Situ Hybridization. 1989. 33 figs. XII, 90 pp.
ISBN 3-540-50761-2

Vol. 144: **Knippers, Rolf; Levine, A. J. (Ed.):** Transforming Proteins of DNA Tumor Viruses. 1989. 85 figs. XIV, 300 pp.
ISBN 3-540-50909-7

Vol. 145: **Oldstone, Michael B. A. (Ed.):** Molecular Mimicry. Cross-Reactivity between Microbes and Host Proteins as a Cause of Autoimmunity. 1989. 28 figs. VII, 141 pp. ISBN 3-540-50929-1

Vol. 146: **Mestecky, Jiri; McGhee, Jerry (Ed.):** New Strategies for Oral Immunization. International Symposium at the University of Alabama at Birmingham and Molecular Engineering Associates, Inc. Birmingham, AL, USA, March 21–22, 1988. 1989. 22 figs. IX, 237 pp. ISBN 3-540-50841-4

Vol. 147: **Vogt, Peter K. (Ed.):** Oncogenes. Selected Reviews. 1989. 8 figs. VII, 172 pp.
ISBN 3-540-51050-8

Vol. 148: **Vogt, Peter K. (Ed.):** Oncogenes and Retroviruses. Selected Reviews. 1989. XII, 134 pp. ISBN 3-540-51051-6

Vol. 149: **Shen-Ong, Grace L. C.; Potter, Michael; Copeland, Neal G. (Ed.):** Mechanisms in Myeloid Tumorigenesis. Workshop at the National Cancer Institute, National Institutes of Health, Bethesda, MD, USA, March 22, 1988. 1989. 42 figs. X, 172 pp.
ISBN 3-540-50968-2

Vol. 150: **Jann, Klaus; Jann, Barbara (Ed.):** Bacterial Capsules. 1989. 33 figs. XII, 176 pp.
ISBN 3-540-51049-4

Vol. 151: **Jann, Klaus; Jann, Barbara (Ed.):** Bacterial Adhesins. 1990. 23 figs. XII, 192 pp. ISBN 3-540-51052-4

Vol. 152: **Bosma, Melvin J.; Phillips, Robert A.; Schuler, Walter (Ed.):** The Scid Mouse. Characterization and Potential Uses. EMBO Workshop held at the Basel Institute for Immunology, Basel, Switzerland, February 20–22, 1989. 1989. 72 figs. XII, 263 pp. ISBN 3-540-51512-7

Vol. 153: **Lambris, John D. (Ed.):** The Third Component of Complement. Chemistry and Biology. 1989. 38 figs. X, 251 pp. ISBN 3-540-51513-5

Vol. 154: **McDougall, James K. (Ed.):** Cytomegaloviruses. 1990. 58 figs. IX, 286 pp. ISBN 3-540-51514-3

Vol. 155: **Kaufmann, Stefan H. E. (Ed.):** T-Cell Paradigms in Parasitic and Bacterial Infections. 1990. 24 figs. IX, 162 pp.
ISBN 3-540-51515-1

Vol. 156: **Dyrberg, Thomas (Ed.):** The Role of Viruses and the Immune System in Diabetes Mellitus. 1990. 15 figs. XI, 142 pp.
ISBN 3-540-51918-1

Vol. 157: **Swanstrom, Ronald; Vogt, Peter K. (Ed.):** Retroviruses. Strategies of Replication. 1990. 40 figs. XII, 260 pp.
ISBN 3-540-51895-9

Vol. 158: **Muzyczka, Nicholas (Ed.):** Viral Expression Vectors. 1990. approx. 20 figs. approx. XII, 190 pp.
ISBN 3-540-52431-2

Vol. 159: **Gray, David; Sprent, Jonathan (Ed.):** Immunological Memory. 1990. 38 figs. XII, 156 pp. ISBN 3-540-51921-1

Vol. 160: **Oldstone, Michael B. A.; Koprowski, Hilary (Ed.):** Retrovirus Infections of the Nervous System. 1990. 16 figs. XII, 176 pp.
ISBN 3-540-51939-4

Vol. 161: **Racaniello, Vincent R. (Ed.):** Piconaviruses. 1990. 12 figs. X, 194 pp.
ISBN 3-540-52429-0

FSC
www.fsc.org
MIX
Papier aus verantwortungsvollen Quellen
Paper from responsible sources
FSC® C105338

If you have any concerns about our products,
you can contact us on
ProductSafety@springernature.com

In case Publisher is established outside the EU,
the EU authorized representative is:
Springer Nature Customer Service Center GmbH
Europaplatz 3, 69115 Heidelberg, Germany

Printed by Libri Plureos GmbH
in Hamburg, Germany